T0315103

The Economics of Microgrids

The Economics of Microgrids

Amin Khodaei
Department of Electrical and Computer Engineering
University of Denver
Denver, CO, USA

Ali Arabnya
Department of Electrical and Computer Engineering
University of Denver
Denver, CO, USA

IEEE PRESS

WILEY

Library of Congress Cataloging-in-Publication Data
Names: Khodaei, Amin, author. | Arabnya, Ali, author.
Title: The economics of microgrids / Amin Khodaei, Ali Arabnya.
Description: Hoboken, New Jersey : Wiley, [2024] | Includes bibliographical
    references and index.
Identifiers: LCCN 2023028414 (print) | LCCN 2023028415 (ebook) | ISBN
    9781394162451 (hardback) | ISBN 9781394162468 (adobe pdf) | ISBN
    9781394162475 (epub)
Subjects: LCSH: Microgrids (Smart power grids)–Costs. | Microgrids (Smart
    power grids)–Economic aspects. | Energy transition.
Classification: LCC TK3105 .K54 2024 (print) | LCC TK3105 (ebook) | DDC
    621.31–dc23/eng/20230825
LC record available at https://lccn.loc.gov/2023028414
LC ebook record available at https://lccn.loc.gov/2023028415

Cover Design: Wiley
Cover Image: © AniGraphics/Getty Images

Set in 9.5/12.5pt STIXTwoText by Straive, Pondicherry, India

Contents

About the Authors

Amin Khodaei, PhD, is a Professor in the Electrical and Computer Engineering department at the University of Denver. His research is focused on the climate crisis, the grid of the future, and advanced technologies to modernize the grid, including artificial intelligence and quantum computing. He has authored/co-authored over 200 peer-reviewed technical papers and has advised over 40 graduate students and postdoctoral associates over the past 10 years. As an active member of the IEEE, he has served as the technical chair of the 2016 and 2018 IEEE PES T&D Conferences and the technical chair of the 2022 IEEE PES General Meeting. He is a Senior Member of IEEE's Power & Energy Society and holds a PhD degree in Electrical Engineering from the Illinois Institute of Technology.

Ali Arabnya, PhD, (also known as Ali Arab) is a Research Professor of Electrical and Computer Engineering at the University of Denver. Previously, he was a consultant climate economist with The World Bank in Washington, DC. Prior to that, he served as Data and Analytics Manager with the Risk and Capital Management practice of Protiviti, a global management consulting firm in New York City. He works at the interface of engineering, finance, and policy in addressing climate change mitigation and adaptation in the power and energy sector. His international professional experience includes North America, the Middle East, and Southeast Asia regions. He is a Senior Member of IEEE's Power and Energy Society and holds a PhD degree in Industrial Engineering from the University of Houston.

Acknowledgments

Several chapters of this book were written based on valuable contributions made by the following co-authors:

Mohammad Shahidehpour, PhD
Sina Parhizi, PhD
Alireza Majzoobi, PhD
Hossein Lotfi, PhD
Mohsen Mahoor, PhD

Acronyms

AC	Alternative current
ANFIS	Adaptive network-based fuzzy inference system
BCR	Benefit-cost ratio
CAGR	Compound annual growth rate
CAIDI	Customer average interruption duration index
CAIFI	Customer average interruption frequency index
CapEx	Capital expenditure
CBM	Condition-based maintenance
DC	Direct current
DES	Distributed energy system
DER	Distributed energy resource
DG	Distributed generation
DMO	Distribution market operator
DNO	Distribution network operator
DSO	Distribution system operator
DSPP	Distributed system platform provider
EDC	Electric distribution company
EMS	Energy management system
ENS	Energy not supplied
ESS	Energy storage system
EV	Electric vehicle
GENCO	Generation company
GHG	Greenhouse gas
HMI	Human–machine interface
IDSO	Independent distribution system operator
ISO	Independent system operator
LC	Local controllers
LDC	Load duration curve
LMP	Locational marginal price

MAE	Mean absolute error
MARR	Minimum acceptable rate of return
MAS	Multi-agent system
MIP	Mixed-integer programming
MILP	Mixed-integer linear program
MLP	Multi-layer perceptron
MMC	Microgrid master controller
NPV	Net present value
O&M	Operations and maintenance
PCC	Point of common coupling
PDC	Price duration curve
PV	Photovoltaics
RBF	Radial basis function
RPS	Renewable portfolio standard
RUL	Remaining useful life
SAIDI	System average interruption duration index
SAIFI	System average interruption frequency index
SCUC	Security-constrained unit commitment
SOC	State of charge
TRANSCO	Transmission company
T&D	Transmission and distribution
UC	Unit commitment
VOLL	Value of lost load
VVC	Volt-VAR control

1

Fundamentals of Microgrids

1.1 Introduction to Microgrids

The earliest idea of the microgrid dates back to 1882, when Thomas Edison built Pearl Street Station, the world's first commercial central power plant, in the Financial District of Manhattan in New York City. Edison's company installed fifty microgrids in four years. At that time, centrally controlled and operated utility grids were not yet formed. With the utility grid subsequently utilizing large, centralized power plants, which benefited from the *economies of scale*, and significantly increased transmission connections for reliability purposes, the electric grid turned into a monopolistic market structure by connecting isolated microgrids, leading these microgrids to be faded away. After more than a century, however, this concept has been revisited, and modern microgrids have gained significant traction, which is driven in part by the need for higher reliability and power quality, higher resilience against disruptive events through grid decentralization, advancements in power electronics, distributed generation (DG), energy storage technologies, and the rise of *prosumers* – i.e. the electricity customers that both consume and produce electric power [1].

While the concept is still evolving, the modern microgrid is defined as "a group of interconnected loads and distributed energy resources (DERs) with clearly defined electrical boundaries that acts as a single controllable entity with respect to the grid and can connect and disconnect from the grid to enable it to operate in both grid-connected or island [*sic*] modes," according to the U.S. Department of Energy [2]. Based on this definition, DER installations can be considered as a microgrid if it comprised three distinct characteristics, as follows: (i) they have electrical boundaries that are clearly defined; (ii) there exists a master controller to control and operate DERs and loads as a single controllable entity; and (iii) the installed generation capacity exceeds the peak critical load, thus it can be disconnected from the utility grid (the islanded mode) and seamlessly supply local critical

The Economics of Microgrids, First Edition. Amin Khodaei and Ali Arabnya.
© 2024 The Institute of Electrical and Electronics Engineers, Inc.
Published 2024 by John Wiley & Sons, Inc.

loads. These characteristics further characterize microgrids as small-scale power systems that self-supply and have islanding capability, which can generate, distribute, and regulate the flow of electricity to local customers.

Microgrids are more than just backup generation units. Backup generation units have existed for quite some time to provide a temporary supply of electricity to local loads when the supply of electricity from the utility grid is interrupted. Microgrids, however, provide a wider range of functions and are significantly more flexible than backup generation units. The main components of microgrids include loads, DERs, master controllers, smart switches, protective devices, as well as communication, control, and automation systems. Microgrid loads are commonly categorized into two types: fixed and flexible (also known as adjustable or responsive) loads. Fixed loads cannot be altered and must be satisfied under normal operating conditions, while flexible loads are responsive to controlling signals. Flexible loads can be curtailed (curtailable loads) or deferred (shiftable loads) in response to economic incentives or islanding requirements. DERs consist of DG units and distributed energy storage systems (ESS), which can be installed at electric utility facilities and/or electricity consumers' premises. Microgrid DGs are either dispatchable or non-dispatchable. Dispatchable units can be controlled by the microgrid master controller and are subject to technical constraints depending on the type of unit, such as capacity limits, ramping limits, minimum on/off time limits, and fuel and emission limits. Non-dispatchable units, on the contrary, cannot be controlled by the microgrid master controller since the input source is uncontrollable. Non-dispatchable units are mainly renewable DGs (typically solar and wind), which produce volatile (generation is fluctuating in different time scales) and intermittent (generation is not always available) output power. These characteristics negatively impact the non-dispatchable unit generations and increase the forecast error; therefore, these units are generally stabilized with ESS. The primary application of ESS is to coordinate with DGs to ensure the microgrid's generation adequacy. They can also be used for energy arbitrage, where the stored energy at low-price hours is generated back to the microgrid when the market price is high. The ESS also plays a major role in microgrid islanding applications. Smart switches and protective devices manage the connection between DERs and loads in the microgrid by connecting/disconnecting distribution lines. When there is a fault in a part of the microgrid, smart switches and protective devices disconnect the faulty area and reroute the power, preventing the fault from propagating in the microgrid. The switch at the point of common coupling (PCC) performs microgrid islanding by disconnecting the microgrid from the main grid. The microgrid scheduling in grid-connected and islanded modes is performed by the microgrid master controller based on economic and security considerations. The master controller determines the microgrid's interaction with the utility grid, the decision to switch between interconnected and islanded modes, and optimal operations of local resources. Communications, control, and automation systems are also used to implement these control actions and to ensure constant, effective, and reliable interaction among microgrid components.

Microgrids offer significant benefits to the customers and the utility grid at system level, as follows: (i) improved reliability by introducing self-healing at the local distribution network; (ii) higher power quality by managing local loads; (iii) reduction in carbon emission by the diversification of energy sources; (iv) economical operations by reducing transmission and distribution (T&D) costs and utilization of less costly renewable DGs; and (v) offering energy efficiency by responding to real-time market prices. The islanding capability is the most salient feature of a microgrid, which is enabled by using switches at the PCC and allows the microgrid to be disconnected from the utility grid in case of upstream disturbances or voltage fluctuations. During utility grid disturbances, the microgrid is transferred from the grid-connected to the islanded mode, and a reliable and uninterrupted supply of consumer loads is offered by local DERs. The islanded microgrid would be resynchronized with the utility grid once the disturbance is eliminated [3, 4].

Microgrids, as enablers of DG integration, improved grid performance, and the green energy economy, have been deployed on a large scale over the past decade and are expected to continue their growth for the foreseeable future. The installed microgrid capacity in 2012 was estimated at 1.1 GW, while as of May 2021, there had been over 460 operational microgrids in the United States that provided a total of 3.1 gigawatts of reliable electricity. The global microgrid market size is estimated to grow from USD 24.6 billion in 2021 to USD 42.3 billion by 2026 at a compound annual growth rate (CAGR) of 11.4% during this period. The expected growth is primarily driven by the decarbonization trend, demand for higher grid resilience, and economies of electrification in rural and remote areas [5].

1.2 Distributed Energy Resources for Microgrids

DERs are small-scale energy resources, which can be placed at utility facilities or at customers' premises to provide a local supply of electricity. DERs can fundamentally change both energy mix and structure of the current energy generation systems – in which the electricity is predominantly generated at large-scale, non-renewable, centralized power plants and transmitted over long distances through "less-than-efficient," high-voltage transmission lines to reach the electricity demand load areas. DER technologies can further provide power to remote locations where required T&D facilities are not available or are costly to build. Moreover, DERs offer a low construction and deployment time compared to large generators and T&D facilities. A comprehensive review of DERs and current practices in microgrids as well as the interaction problems arising from integration of various DERs in a microgrid can be found in [6, 7], respectively. As discussed in detail in [8], DERs include a variety of technologies. Two widely used DERs include renewable DGs and ESS. There has been an increasing traction for the utilization of renewable DGs, such as wind and solar energy resources, in recent years. That is primarily due to the regulatory mandates, reduced cost of renewable generation, and policy incentives for decarbonization and reduction of greenhouse gas (GHG) emissions to address climate change. Renewable DGs are primarily dependent on meteorological factors, resulting in high levels of variability and uncertainty in their power generation share. That is one of the challenges that need to be overcome in order to enable broader integration of renewable energy resources. Sunlight is the origin of most renewable DGs either directly (e.g. solar energy), or indirectly (e.g. wind, hydroelectric, and biomass energy resources). Sunlight is directly converted to solar energy using solar panels. Wind and hydroelectric power are the results of differential heating of the earth's surface. Biomass energy is the sunlight energy stored in plants. There are also some other types of energy not driven by the sun, such as geothermal energy, whose origin is the internal heat in the Earth, as well as ocean energy, which comes from tides and winds. Renewable DGs offer several benefits including sustainability, being emission-free, and benefiting from almost ubiquitous primary sources of energy [9, 10]. As detailed in [11], there are numerous policies and regulations rolled out in a number of states within the United States to support investments in renewable DGs. Some of the examples include renewable portfolio standards, public benefit funds for renewable energy, output-based environmental regulations, interconnection standards, net metering, feed-in tariffs, property-assessed clean energy, and other financial incentives. Based on renewable portfolio standards, all electricity providers should provide a specific amount of electric power using renewable DGs. Public benefit funds are obtained by levying small taxes on electricity rates. Output-based environmental regulations (such as cap-and-trade programs) ordain emission limits in order to encourage electric producers to increase efficiency and control air pollution. Interconnection standards are technical requirements, which should be met by electricity providers that want to connect renewable DGs to the grid. These standards determine how electric utilities in a jurisdiction would treat renewable DGs. Net metering rules are used to compensate for the power generation by prosumers. For instance, if the local power generated by a customer is more than their load, the surplus power is sold back to the utility grid, and on the other hand, if the generated power is not sufficient to supply loads, they use electricity from the utility grid. This procedure requires accurate metering of the electricity demand from consumers. Feed-in tariff is a policy incentive to encourage renewable energy development, which requires electric utilities to make long-term payments for the power fed into the grid by renewable energy developers. The payments may comprise both electricity sales and payments for renewable energy certificates. Feed-in tariff policies can economically incentivize a rapid development of renewable DGs. Implementing feed-in tariffs has been a successful experience to meet economic development and renewable energy targets around the world. Based on property-assessed clean energy policies, the cost of renewable energy installations or increasing energy efficiency is refunded to residential properties instead of individual borrowers. As a result, property owners would be encouraged to invest in renewable energy deployment on their premises.

The variable nature of renewable DGs in microgrids necessitates the presence of an energy source to compensate for their fluctuations. The islanding events in microgrids and the need for a power supply to ensure seamless transfer to the islanded mode also make the case for integration of ESS in microgrids. ESS enhances flexibility in power generation, delivery, and consumption. It provides utility grids with several benefits and large cost savings. Large-scale ESS increases the efficiency of utility grids, which means lower operations costs, reduced emissions, and increased reliability. Considering the increasing penetration of renewable DGs and their intermittent nature, the application of ESS has significantly increased in recent years. For example, wind farms or solar photovoltaics (PVs) generate power when the wind is blowing or the sun is shining. Accordingly, the employment of ESS allows the utility grid to store energy when it is more than the amount required to meet the demand and supply loads in peak hours. Therefore, this technology enables variable generation resources to continue their power generation even in the absence of wind and sunlight, which means providing electric utilities with continuous and reliable power. Storing energy from various resources to economically serve shifting loads based on electricity prices and to serve non-shiftable loads during peak hours is one of the several applications of ESS. The deployment of ESSs can improve power quality via frequency regulation, benefit electric producers by allowing them to generate power when it is most efficient and least expensive, provide critical loads with a continuous source of power, and help customers during emergencies such as power outages due to climate shocks and natural disasters, equipment failures, or malicious cyber-physical attacks. As discussed in [12], benefits can be in the form of either avoided costs or additional revenue received by the operator. Based on this concept, if an ESS is used such that there is no need for generation equipment, the economic benefit of this ESS includes but is not limited to the avoided cost of generation. For the ESS owner, benefits from additional revenue can be realized from selling surplus energy and other services. For an electricity end-user deploying ESS for reducing electricity bills, the economic benefits can be realized from lower cost of energy [13, 14]. In [15] microgrids were classified based on their value proposition into three types, as follows: (i) reliability; (ii) energy arbitrage; and (iii) power quality.

It was shown that as energy source of inverter-based microgrids responds slowly, ESS is only necessary for inverter-based microgrids – with a most critical load designed to have a power quality higher than the utility grid – and is optional for other types of microgrids. Reference [16] demonstrates that ESSs deployed in microgrids can perform the tasks of active power balancing and voltage regulation at the same time. In the grid-connected mode, ESS may ensure load leveling and reduce the power exchange with the network, which makes the system's operation more efficient and flexible. In addition, ESS may enhance DER penetration and contribute to better quality of energy delivery to customers.

A wide range of DERs are not suitable to be directly connected to the microgrid network. Therefore, power electronic interfaces are required to enhance/enable their integration [6]. Examples are PV cells and ESS, which generate DC power, or wind turbines that need improvements in generated power quality and frequency. Although power electronic devices would enhance integration and controllability of these resources, they can also bring new challenges regarding control and protection. In an islanded microgrid, rotating generators can serve the role of a voltage source and manage the grid frequency, but in their absence, the power electronic converters are needed to behave as voltage sources. In the grid-connected mode, the converters function as current sources feeding the microgrid. In addition to enabling an efficient connection, power electronics devices are capable of providing additional benefits to microgrids. Power electronic interfaces can improve the power quality of customers by improving harmonics and providing extremely fast switching times for sensitive loads. Power electronics can also provide benefits to the connected utility grid by providing reactive power control and voltage regulation at the distributed energy system connection point. A useful feature of a power electronic interface is the ability to reduce or eliminate fault current contributions from distributed energy systems, thereby allowing negligible impacts on protection coordination. Finally, power electronic interfaces provide flexibility in operations with various other DERs and can potentially reduce overall interconnection costs through standardization and modularity [17].

1.3 The Role of Microgrid in Power Systems

Microgrids can play an important role in power grids through improved reliability, increased resilience, emission reduction, reduced costs of recurring system upgrades, enhanced energy efficiency and power quality, lowered energy costs, and financial gains through energy arbitrage [18, 19]. In this section, we focus on the most important roles that microgrids can play in power systems, as follows.

1.3.1 Reliability

One of the most important benefits of microgrids is that they improve power supply reliability. Electric utilities constantly monitor customers' reliability levels and perform required system upgrades to improve supply availability and to reach or maintain desired performance. Consumer reliability is typically evaluated in terms of system and customer average interruption frequency and/or duration indices (such as System Average Interruption Frequency Index "SAIFI," System Average Interruption Duration Index "SAIDI," Customer Average Interruption Frequency Index "CAIFI," and Customer Average Interruption Duration Index "CAIDI"). Outage causes such as storms or equipment failure, can impact reliability levels by increasing the average frequency and duration of interruptions; however, when a microgrid is deployed, these metrics can be significantly improved. This is due to the intrinsic intelligence (control and automation systems) of microgrids and the utilization of DERs that allow islanded operation from the utility grid. In particular, since the generation in community microgrids is located in close proximity to consumer loads, it is less prone to be exposed to and affected by grid disturbances and infrastructure issues. Additional flexibility to provide service under these conditions is provided by the ability to adjust loads (e.g. demand response) via building and/or microgrid master controllers. Improved reliability can be translated into economic benefits for consumers and utility due to a reduction in interruption costs and the amount of energy not supplied (ENS). The magnitude of these economic benefits is dependent upon load criticality, the value of lost load (VOLL), and also the availability of other alternatives such as backup generation or automatic load transfer. Microgrid studies associated with reliability can be considered from two perspectives: evaluation and improvement. For example, in the context of microgrid reliability evaluation, studies in [20–26] consider reliability assessment of islanded microgrids with renewable DGs. In the context of microgrid reliability improvement, methods for increasing reliability have been proposed through coupled microgrids [27], adding renewable DGs [28], autonomous customer-driven microgrids [29], efficient operation of DGs [30], and vehicle-to-grid integration [31], among others.

1.3.2 Resiliency

Resiliency refers to the capability of power systems to withstand low-probability-high-impact events by minimizing possible power outages and quickly returning to normal operating state [32]. These events include extreme weather events and natural disasters, such as hurricanes, tornadoes, earthquakes, snowstorms, and floods, as well as manmade disasters such as cyberattacks and malicious physical attacks, among others. Recent climate shocks to the power grid infrastructure in the United States and the potential significant social disruptions have spawned a great deal of debates in the power and energy industry about the value and application of microgrids. If the power system is impacted by these events and critical components are severely damaged (e.g. generating facilities and T&D infrastructure), service may be disrupted for days, weeks, or even a longer period. The impact of these events on consumers can be minimized by decentralizing the grid by deployment of microgrids, which enable the local supply of loads even when the supply of power from the utility grid is not available. Examples of research work in this area include [33–38], among others.

1.3.3 Power Quality

Consumers' need for higher power quality has significantly increased during past decades due to the growing application of voltage-sensitive loads, including a large number and variety of electronic loads and LEDs. Utilities are always seeking efficient ways of improving power quality issues by addressing prevailing concerns stemming from harmonics and voltage. Microgrids provide a quick and efficient answer for addressing power quality needs by enabling local control of frequency, voltage, load, and the rapid response from ESS. For example, power quality improvement through microgrids has been investigated in [39–46].

1.4 Microgrid Technologies

1.4.1 Microgrid Power Management and Control

It is common to use a hierarchical control structure to control microgrids. The hierarchical structure typically consists of three broad layers, as follows: (i) primary control that stabilizes frequency and voltage using droop controllers; (ii) secondary control that compensates the steady state deviations in voltage and frequency caused by the primary control; and (iii) tertiary control that takes into account economic considerations and determines power flow between the microgrid and utility grid to achieve an optimal operation [47–49]. In addition to the control structure, control methods are very important as well. Some renewable DGs such as wind and solar PV have fluctuations and do not generate constant power. As a result, microgrid control can be a complex and difficult process. The study in [50] states that there are two main control methods for microgrids: controller/responder and peer-to-peer control mechanisms. The former is associated with voltage–frequency (V–f) control, while other DGs are associated with

P–Q control to control the active and reactive power to be reached to the planned targets. Peer-to-peer control is associated with frequency–active power (f–P) and voltage–reactive power (V–Q) controls. Both controls (controller/responder and peer-to-peer) have their own advantages and disadvantages.

Two common control architectures for microgrids are centralized and distributed. Standardized procedures and easy implementations are among the advantages of the centralized approach. The study in [51] presents a microgrid central controller with two major functions for distribution systems that include a communication channel with the distribution system operator and the electricity market and exchanging information with the microgrid local controllers (LCs) and processing them. In the centralized control scheme, the central controller makes decisions about the dispatch of all DGs and ESSs according to the objective function and constraints. In microgrids where each DG has its own controller and pursues distinct objectives, distributed control provides premium applicability. The number of transmitted messages between different individual components and the microgrid controller increases as the size of the microgrid increases, necessitating a larger communication bandwidth. Decentralized control can reduce the number of messages and simplify the optimization with special constraints by reducing it into subproblems and solving them locally [52]. One approach to implement distributed control is based on using multi-agent systems (MAS). In this approach, each of the controllable elements in the microgrid, such as inverters, loads, and DGs, have agents associated with them, where the communication and coordination of the agents is governed by the multi-agent theory. MAS includes the microgrid cluster management agent, microgrid control agent, and local agent. The loosely coupled agents forming the MAS are physically or logically dispersed and have a set of distinct characteristics, as follows: (i) their data is distributed; (ii) they have an asynchronous or simultaneous process of computation; (iii) they lack information and capability of problem-solving; and (iv) they interact and cooperate with each other, hence their problem-solving capability can be improved [53].

When a microgrid becomes islanded from the utility grid, the primary control keeps the voltage and frequency stable. However, the voltage and frequency can still divert from their nominal values. In order to retrieve the voltage and frequency to nominal values, a secondary control mechanism should be employed. This secondary control can be the distributed cooperative control. "Cooperative" means that all participants cooperate with each other and act as a single group to reach the common goals [54]. Synchronous generators exhibit a self-stabilizing feature due to their high rotational inertia [55]. Most of the generation units integrated in the microgrid are not classified as synchronous generators and need to mimic the droop characteristic of those generators. When connected to the utility grid, the microgrid voltage will be dictated by the utility grid as it acts as an infinite bus. In the islanded mode, however, voltage control becomes an important and challenging task that requires careful attention. Most of DERs installed in the microgrid generate DC or variable-frequency power that unlike synchronous generators cannot be relied on for frequency regulation in the islanded operation. The high penetration of power-electronically interfaced DGs leads to a low inertia in microgrids. Therefore, proper measures need to be implemented to control frequency in the microgrid [56]. Adaptive control schemes can be used to control the systems with varying or uncertain parameters. As microgrid operating modes can unexpectedly change as a result of disturbances in the utility grid, adaptive control schemes are proposed. When the microgrid power exchange with the utility grid is scheduled, it is necessary to establish a control mechanism so that the actual power flow matches the scheduled values. The control of the power flow between the microgrid and the utility grid has been the main discussion in [57–60].

1.4.2 Microgrid Islanding

The salient feature of a microgrid is its ability to be islanded from the utility grid by upstream switches at the PCC. Islanding can be introduced for economic as well as reliability purposes. During utility grid disturbances, microgrids can transition from the grid-connected to the islanded mode, where a reliable and uninterrupted supply of consumer loads can be provided by local DERs. The microgrid master controller can facilitate optimal operations by maintaining the frequency and voltages within permissible ranges. The islanded microgrid can be resynchronized with the utility grid once the disturbance is eliminated [61–63]. Once the fault is alleviated, the microgrid will be resynchronized with the utility grid. Resynchronization refers to reconnecting the islanded microgrid to the utility grid while ensuring that the microgrid voltage and frequency are synchronized with those of the utility grid [64]. If not ensured, serious damage due to current surges may happen to the microgrid components during the switching process. Although microgrids are infrequently switched to the islanded mode, there could be significant social cost savings and load point reliability enhancements offered by microgrids during major outages.

1.4.3 Microgrid Protection

The unique characteristics of microgrids necessitate changes to the conventional distribution network protection strategies. Connection of DERs, which are normally power electronically interfaced, results in bidirectionality of fault current, reduction in fault current capacity, disruption in fault detection, and protection sensitivity. Furthermore, the dynamic topology of the microgrids due to islanding and sectionalizing necessitates the protection to be able to adapt itself to new conditions. Due to variable microgrid operating conditions and meshed topology of microgrids, it is necessary to use communications to update protection settings [65, 66]. The study in [65] shows that the traditional communication-less protection schemes are not applicable in a meshed microgrid where a fault at one location is indistinguishable from another. In [67], a protection scheme is presented using digital relays with a communication network for the protection of the microgrid, relying primarily on differential protection based on sampling the current waveform. IEC 61850 is an international standard for substation automation and a part of the International Electrotechnical Commission's Technical Committee 57 (TC57) architecture for electric power systems. These standards will result in very significant improvements in both costs and the performance of utility grids. They are based on abstracting definition of the data items and the services, or, in other words, creating data items/objects and services that are independent of any underlying protocols. The abstract definitions then allow mapping of the data objects and services to any other protocol that can meet the data and service requirements [68]. Due to the existence of different levels of fault current in microgrids, new protective schemes need to be developed that can monitor changes in the microgrid and calculate the operating conditions at any given time. Logical nodes available in IEC 61850 and IEC 61850-7-420 communication standards are used to design such versatile schemes in microgrids [69].

1.4.4 Microgrid Communications and Human–Machine Interface (HMI)

The role of communication systems in the microgrid is to provide a means to exchange data and monitor various elements for control and protection purposes. In a centrally controlled microgrid, the communication network is necessary to communicate control signals to the microgrid components. In a microgrid with distributed control, the communication network enables each component to communicate with other components in the microgrid, decide on its operation, and further reach predefined objectives [51]. Communications within the microgrid are necessary to enable rapid fault clearing and increase efficiency in islanding incidences. The communications structure for microgrids includes a three-layer, hierarchical architecture, as follows: (i) the top layer hosts the energy management system (EMS) that controls the overall operations of the microgrids in both interconnected and islanded modes; (ii) the middle layer is comprised of LCs that regulate the microgrid operations and its interactions with the main grid; and (iii) the bottom layer, which consists of IoT devices (e.g. smart meters, fault recorders, and protective relays), that continuously monitors, records, and transmits the stream of sensed data [70]. An important building block of the EMS is constituted of human–machine interface (HMI), which includes hardware or software through which the microgrid operators interact with the microgrid controller. HMI facilitates on-demand microgrid monitoring and control on a real-time basis through a two-way communication network. On the system operator side, that includes visualizing operations, archiving the collected data, and processing command information. On the customer side, it includes enabling customers to actively participate in and interact with the EMS [51]. An example of a system design for a microgrid EMS that includes details of HMI can be found in [71].

1.5 Overview

This book discusses the engineering economics of microgrids by covering the economic decision-making processes involved in the system design and operations of these systems. The remainder of the book is organized as follows:

- Chapter 2: *Microgrid Operations Economics* introduces the economics of operations management for microgrids and shows how islanding can economically impact their operations.
- Chapter 3: *Resilience Economics in Microgrids* discusses the economics of resilience in microgrids for optimal operations of these systems during outages and power disturbances.
- Chapter 4: *Community Microgrid Operations Management* presents economic operations scheduling models for community microgrids without compromising the privacy of the users.
- Chapter 5: *Provisional Microgrids for Renewable Energy Integration* introduces the economic decision-making framework for operations of a novel class of microgrids, namely provisional microgrids, that are important enablers of renewable energy integration in power systems.
- Chapter 6: *Engineering Economics of Microgrid Investments* presents an analytical framework on investment decisions and capital expenditure analysis required for the economic assessment of microgrid projects.
- Chapter 7: *Microgrid Planning Under Uncertainty* presents an economic operations management model that incorporates the uncertainties associated with the prediction of the loads and the market price.
- Chapter 8: *Microgrid Expansion Planning* discusses economic concepts and models for minimizing microgrids' operations costs, including the cost of local generation resources and energy purchases from the main grid to supply local loads.
- Chapter 9: *Microgrids for Asset Management in Power Systems* presents an asset management strategy for distribution networks that incorporates microgrids to maximize the remaining useful life of critical assets in the grid.

- Chapter 10: *Dynamics of Microgrids in Distribution Network Flexibility* presents an economic model for using microgrids to support electricity distribution networks by improving their flexibility and eliminating costly investment alternatives.
- Chapter 11: *Microgrid Operations Under Electricity Market Dynamics* introduces an economic decision-making model that incorporates the impacts of electricity markets on microgrid operations and planning.

Each chapter of this book is designed to stand on its own and has its own introduction, nomenclature, bibliography, and acronyms.

References

1 Department of Energy Office of Electricity Delivery and Energy Reliability, "Summary Report: 2012 DOE Microgrid Workshop," 2012. [Online]. Available: http://energy.gov/sites/prod/files/2012 Microgrid Workshop Report 09102012.pdf. [Accessed: 30-Dec-2022].

2 Herman, D., "Investigation of the Technical and Economic Feasibility of Micro-Grid Based Power Systems," vol. 2, pp. 1, Palo Alto, CA: Electric Power Research Institute, 2001.

3 "What Are the Benefits of the Smart Microgrid Approach? | Galvin Electricity Initiative." [Online]. Available: http://www.galvinpower.org/resources/microgrid-hub/smart-microgrids-faq/benfits. [Accessed: 13-Feb-2015].

4 "Microgrids—Benefits, Models, Barriers and Suggested Policy Initiatives for the Commonwealth of Massachusetts | MassCEC." [Online]. Available: http://www.masscec.com/content/microgrids-%E2%80%93-benefits-models-barriers-and-suggested-policy-initiatives-commonwealth. [Accessed: 13-Feb-2015].

5 A. Arab and A. Khodaei, "An Economic Evaluation Framework for Sustainable Community Microgrids," in *CIGRE US National Committee, Grid of the Future Symposium*, Chicago, IL, 2022.

6 H. Jiayi, J. Chuanwen, and X. Rong, "A review on distributed energy resources and MicroGrid," *Renew. Sustain. Energy Rev.*, vol. 12, no. 9, pp. 2472–2483, Dec. 2008.

7 Y. Zoka, H. Sasaki, N. Yorino, K. Kawahara, and C. C. Liu, "An interaction problem of distributed generators installed in a MicroGrid," in *2004 IEEE International Conference on Electric Utility Deregulation, Restructuring and Power Technologies. Proceedings*, vol. 2, pp. 795–799, 2004.

8 S. Chowdhury and P. Crossley, Microgrids and Active Distribution Networks, Institution of Engineering and Technology, 2009.

9 "Why Is Renewable Energy Important? [Online]. Available: http://www.renewableenergyworld.com/rea/tech/home. [Accessed: 13-Feb-2015].

10 "Renewable Energy, Forms and Types of Renewable Energy." [Online]. Available: http://www.altenergy.org/renewables/renewables.html. [Accessed: 13-Feb-2015].

11 US Environmental Protection Agency, "State and Local Climate and Energy Program." [Online]. Available: http://www.epa.gov/statelocalclimate/state/topics/renewable.html. [Accessed: 13-Feb-2015].

12 S. Parhizi, H. Lotfi, A. Khodaei, and S. Bahramirad, "State of the art in research on microgrids: a review," *IEEE Access*, vol. 3, pp. 890–925, 2015.

13 X. Tan, Q. Li, and H. Wang, "Advances and trends of energy storage technology in microgrid," *Int. J. Electr. Power Energy Syst.*, vol. 44, no. 1, pp. 179–191, Jan. 2013.

14 Z. Xu, X. Guan, Q.-S. Jia, J. Wu, D. Wang, and S. Chen, "Performance analysis and comparison on energy storage devices for smart building energy management," *IEEE Trans. Smart Grid*, vol. 3, no. 4, pp. 2136–2147, Dec. 2012.

15 A. D. Paquette and D. M. Divan, "Design considerations for microgrids with energy storage," in *2012 IEEE Energy Conversion Congress and Exposition (ECCE)*, pp. 1966–1973, 2012.

16 R. Pawelek, I. Wasiak, P. Gburczyk, and R. Mienski, "Study on operation of energy storage in electrical power microgrid—modeling and simulation," in *Proceedings of 14th International Conference on Harmonics and Quality of Power—ICHQP 2010*, 2010.

17 B. Kroposki, C. Pink, R. DeBlasio, H. Thomas, M. Simões, and P.K. Sen, "Benefits of power electronic interfaces for distributed energy systems," *IEEE Trans. Energy Convers.*, vol. 25, no. 3, pp. 901–908, Sept. 2010.

18 S. Bahramirad, A. Khodaei, J. Svachula, and J. R. Aguero, "Building resilient integrated grids: one neighborhood at a time," *IEEE Electrif. Mag.*, vol. 3, no. 1, pp. 48–55, Mar. 2015.

19 A. D. Paquette and D. M. Divan, "Providing improved power quality in microgrids: difficulties in competing with existing power-quality solutions," *IEEE Ind. Appl. Mag.*, vol. 20, no. 5, pp. 34–43, Sept. 2014.

20 H. Daneshi and H. Khorashadi-Zadeh, "Microgrid energy management system: a study of reliability and economic issues," in *2012 IEEE Power and Energy Society General Meeting*, San Diego, CA, pp. 1–5, 2012.

21 B. Falahati and A. Kargarian, "Timeframe capacity factor reliability model for isolated microgrids with renewable energy resources," in *2012 IEEE Power and Energy Society General Meeting*, San Diego, CA, pp. 1–8, 2012.

22 S. Kennedy, "Reliability evaluation of islanded microgrids with stochastic distributed generation," in *2009 IEEE Power & Energy Society General Meeting*, Calgary, AB, pp. 1–8, 2009.

23 A. K. Basu, S. Chowdhury, and S. P. Chowdhury, "Distributed energy resource capacity adequacy assessment for PQR enhancement of CHP micro-grid," in *IEEE PES General Meeting*, Minneapolis, MN, pp. 1–5, 2010.

24 R. Yokoyama, T. Niimura, and N. Saito, "Modeling and evaluation of supply reliability of microgrids including PV and wind power," in *2008 IEEE Power and Energy Society General Meeting—Conversion and Delivery of Electrical Energy in the 21st Century*, Pittsburgh, PA, pp. 1–5, 2008.

25 S. Wang, Z. Li, L. Wu, M. Shahidehpour, and Z. Li, "New metrics for assessing the reliability and economics of microgrids in distribution system," *IEEE Trans. Power Syst.*, vol. 28, no. 3, pp. 2852–2861, Aug. 2013.

26 H. E. Farag, M. M. A. Abdelaziz, and E. F. El-Saadany, "Voltage and reactive power impacts on successful operation of islanded microgrids," *IEEE Trans. Power Syst.*, vol. 28, no. 2, pp. 1–1, Nov. 2012.

27 R. H. Lasseter, "Smart distribution: coupled microgrids," *Proc. IEEE*, vol. 99, no. 6, pp. 1074–1082, June 2011.

28 H. E. Brown, S. Suryanarayanan, S. A. Natarajan, and S. Rajopadhye, "Improving reliability of islanded distribution systems with distributed renewable energy resources," *IEEE Trans. Smart Grid*, vol. 3, no. 4, pp. 2028–2038, Dec. 2012.

29 J. Mitra and S. J. Ranade, "Power system hardening through autonomous, customer-driven microgrids," in *2007 IEEE Power Engineering Society General Meeting*, Tampa, FL, 2007.

30 K.-H. Kim, S.-B. Rhee, K.-B. Song, and K. Y. Lee "An efficient operation of a micro grid using heuristic optimization techniques: harmony search algorithm, PSO, and GA," in *2012 IEEE Power and Energy Society General Meeting*, San Diego, CA, pp. 1–6, 2012.

31 A. J. Markel, "Simulation and analysis of vehicle-to-Grid operations in microgrid," in *2012 IEEE Power and Energy Society General Meeting*, San Diego, CA, pp. 1–5, 2012.

32 A. Khodaei, "Resiliency-oriented microgrid optimal scheduling," *IEEE Trans. Smart Grid*, vol. 5, no. 4, pp. 1584–1591, July 2014.

33 X. Xu, J. Mitra, N. Cai, and L. Mou, "Planning of reliable microgrids in the presence of random and catastrophic events," *Int. Trans. Electr. Energy Syst.*, vol. 24, pp. 1151–1167, July 2013.

34 S. Cano-Andrade, M. R. von Spakovsky, A. Fuentes, C. Lo Prete, B. F. Hobbs, and L. Mili, "Multi-objective optimization for the sustainable-resilient synthesis/design/operation of a power network coupled to distributed power producers via microgrids," in ASME International Mechanical Engineering Congress and Exposition, vol. 45226, American Society of Mechanical Engineers, 2012.

35 R. Arghandeh, M. Pipattanasomporn, and S. Rahman, "Flywheel energy storage systems for ride-through applications in a facility microgrid," *IEEE Trans. Smart Grid*, vol. 3, no. 4, pp. 1955–1962, Dec. 2012.

36 A. Kwasinski, V. Krishnamurthy, J. Song, and R. Sharma, "Availability evaluation of micro-grids for resistant power supply during natural disasters," *IEEE Trans. Smart Grid*, vol. 3, no. 4, pp. 2007–2018, Dec. 2012.

37 F. O. Resende, N. J. Gil, and J. A. P. Lopes, "Service restoration on distribution systems using multi-MicroGrids," *Eur. Trans. Electr. Power*, vol. 21, no. 2, pp. 1327–1342, Mar. 2011.

38 J. Hurtt and L. Mili, "Residential microgrid model for disaster recovery operations," in *2013 IEEE Grenoble Conference*, Grenoble, France, pp. 1–6, 2013.

39 R. M. Kamel, A. Chaouachi, and K. Nagasaka, "Wind power smoothing using fuzzy logic pitch controller and energy capacitor system for improvement micro-grid performance in islanding mode," *Energy*, vol. 35, no. 5, pp. 2119–2129, May 2010.

40 Y. Ito, Y. Zhongqing, and H. Akagi, "DC microgrid based distribution power generation system," in *The 4th International Power Electronics and Motion Control Conference*, Xi'an, China, vol. 3. pp. 1740–1745, 2004.

41 S. Chakraborty, M. D. Weiss, and M. G. Simoes, "Distributed intelligent energy management system for a single-phase high-frequency AC microgrid," *IEEE Trans. Ind. Electron.*, vol. 54, no. 1, pp. 97–109, Feb. 2007.

42 F. Wang, J. L. Duarte, and M. A. M. Hendrix, "Grid-interfacing converter systems with enhanced voltage quality for microgrid application—concept and implementation," *IEEE Trans. Power Electron.*, vol. 26, no. 12, pp. 3501–3513, Dec. 2011.

43 Y. W. Li, D. M. Vilathgamuwa, and P. C. Loh, "A grid-interfacing power quality compensator for three-phase three-wire microgrid applications," *IEEE Trans. Power Electron.*, vol. 21, no. 4, pp. 1021–1031, July 2006.

44 K. T. Tan, P. L. So, Y. C. Chu, and M. Z. Q. Chen, "A flexible AC distribution system device for a microgrid," *IEEE Trans. Energy Convers.*, vol. 28, no. 3, pp. 601–610, Sept. 2013.

45 Y. Li, D. M. Vilathgamuwa, and P. C. Loh, "Microgrid power quality enhancement using a three-phase four-wire grid-interfacing compensator," *IEEE Trans. Ind. Appl.*, vol. 41, no. 6, pp. 1707–1719, Nov. 2005.

46 M. Illindala and G. Venkataramanan, "Frequency/sequence selective filters for power quality improvement in a microgrid," *IEEE Trans. Smart Grid*, vol. 3, no. 4, pp. 2039–2047, Dec. 2012.

47 J.W. Simpson-Porco and F. Bullo, "Breaking the Hierarchy: Distributed Control & Economic Optimality in Microgrids," Canada: National Science Foundation NSF CNS-1135819 and National Science and Engineering Research Council, Jan. 2014.

48 A. Bidram and A. Davoudi, "Hierarchical structure of microgrids control system," *IEEE Trans. Smart Grid*, vol. 3, no. 4, pp. 1963–1976, Dec. 2012.

49 J. M. Guerrero, J. C. Vasquez, J. Matas, L. G. de Vicuna, and M. Castilla, "Hierarchical control of droop-controlled AC and DC microgrids—a general approach toward standardization," *IEEE Trans. Ind. Electron.*, vol. 58, no. 1, pp. 158–172, Jan. 2011.

50 Y. Che and J. Chen, "Research on design and control of microgrid system," *Prz. Elektrotech.*, no. 5, pp. 83–86, Jan. 2012.

51 W. Su and J. Wang, "Energy management systems in microgrid operations," *Electr. J.*, vol. 25, no. 8, pp. 45–60, Oct. 2012.

52 A. Dimeas, A. Tsikalakis, G. Kariniotakis, and G. Korres, 2013 "Microgrid: Architectures and Control," Hatziargyriou, Nikos, ed., John Wiley & Sons, pp. 26–76, 2014.

53 G. Zheng and N. Li, "Multi-agent based control system for multi-microgrids," in *2010 International Conference on Computational Intelligence and Software Engineering*, Wuhan, China, pp. 1–4, 2010.

54 A. Bidram, A. Davoudi, F.L. Lewis, and J.M. Guerrero, "Distributed cooperative secondary control of microgrids using feedback linearization," *IEEE Trans. Power Syst.*, vol. 28, no. 3, pp. 3462–3470, Aug. 2013.

55 J. Driesen and K. Visscher, "Virtual synchronous generators," in *2008 IEEE Power and Energy Society General Meeting—Conversion and Delivery of Electrical Energy in the 21st Century*, Pittsburgh, PA, pp. 1–3, 2008.

56 S. Riverso, F. Sarzo, and G. Ferrari-Trecate, "Plug-and-play voltage and frequency control of islanded microgrids with meshed topology," *IEEE Trans. Smart Grid*, vol. 6, no. 99, pp. 1, May 2015.

57 T. L. Vandoorn, J. D. M. De Kooning, B. Meersman, J. M. Guerrero, and L. Vandevelde, "Voltage-based control of a smart transformer in a microgrid," *IEEE Trans. Ind. Electron.*, vol. 60, no. 4, pp. 1291–1305, Apr. 2013.

58 R. Majumder, A. Ghosh, G. Ledwich, and F. Zare, "Power management and power flow control with back-to-back converters in a utility connected microgrid," *IEEE Trans. Power Syst.*, vol. 25, no. 2, pp. 821–834, May 2010.

59 M. G. Molina and P. E. Mercado, "Power flow stabilization and control of microgrid with wind generation by superconducting magnetic energy storage," *IEEE Trans. Power Electron.*, vol. 26, no. 3, pp. 910–922, Mar. 2011.

60 J. Gervasoni, M. G. Molina, and P. E. Mercado, "Stabilization and control of tie-line power flow of microgrid including wind generation by distributed energy storage," *Int. J. Hydrogen Energy*, vol. 35, no. 11, pp. 5827–5833, June 2010.

61 F. Katiraei and M. R. Iravani, "Power management strategies for a microgrid with multiple distributed generation units," *IEEE Trans. Power Syst.*, vol. 21, no. 4, pp. 1821–1831, Nov. 2006.

62 C. Hou, X. Hu, and D. Hui, "Hierarchical control techniques applied in micro-grid," in *2010 International Conference on Power System Technology*, Zhejiang, China, 2010.

63 A. G. Tsikalakis and N. D. Hatziargyriou, "Centralized control for optimizing microgrids operation," *IEEE Trans. Energy Convers.*, vol. 23, no. 1, pp. 241–248, Mar. 2008.

64 J. Hossain and A. Mahmud, "Renewable Energy Integration: Challenges and Solutions," Springer Science & Business Media, 2014.

65 A. Prasai, Y. Du, A. Paquette, E. Buck, R. Harley, and D. Divan, "Protection of meshed microgrids with communication overlay," in *2010 IEEE Energy Conversion Congress and Exposition*, Atlanta, GA, pp. 64–71, 2010.

66 T. S. Ustun, C. Ozansoy, and A. Zayegh, "Simulation of communication infrastructure of a centralized microgrid protection system based on IEC 61850-7-420," in *2012 IEEE Third International Conference on Smart Grid Communications (SmartGridComm)*, Tainan, Taiwan, pp. 492–497, 2012.

67 E. Sortomme, S. S. Venkata, and J. Mitra, "Microgrid protection using communication-assisted digital relays," *IEEE Trans. Power Deliv.*, vol. 25, no. 4, pp. 2789–2796, Oct. 2010.

68 R. E. Mackiewicz, "Overview of IEC 61850 and benefits," in *2006 IEEE Power Engineering Society General Meeting*, Atlanta, GA, p. 8, 2006.

69 T. S. Ustun, C. Ozansoy, and A. Zayegh, "Modeling of a centralized microgrid protection system and distributed energy resources according to IEC 61850-7-420," *IEEE Trans. Power Syst.*, vol. 27, no. 3, pp. 1560–1567, Aug. 2012.

70 S. Kumar, S. Islam, and A. Jolfaei, "Microgrid communications—protocols and standards," in Variability, Scalability and Stability of Microgrids, Institution of Engineering and Technology, pp. 291–326, 2019.

71 W. Shi, E. K. Lee, D. Yao, R. Huang, C. C. Chu, and R. Gadh, "Evaluating microgrid management and control with an implementable energy management system," in *IEEE International Conference on Smart Grid Communications*, Venice, Italy, 2014.

2

Microgrid Operations Economics

2.1 Fundamentals of Microgrid Operations Economics

The microgrid is economically operated in grid-connected mode; however, sufficient capacity should always be available in case the microgrid is required to switch to the islanded mode. The microgrid is islanded from the main grid using upstream switches at the point of common coupling (PCC), and the microgrid load is fully supplied using local resources [1–3].

The microgrid scheduling in grid-connected and islanded mode is performed by the microgrid master controller based on security and economic considerations. The microgrid optimal scheduling performed by the microgrid master controller is considerably different from the unit commitment (UC) problem solved by an independent system operator (ISO) for the main grid. Variable generation resources and energy storage systems have major roles in microgrid operation due to their considerable size compared to local loads. In addition, generation resources are close to load premises, and power is transmitted over medium- or low-voltage distribution networks; hence, congestion would not be an issue in power transfer. A high percentage of local loads could also be responsive to price variations, which makes the microgrid load/generation balance more flexible. Finally, the connection to the main grid in grid-connected mode, which represents the main grid as an infinite bus with unlimited power supply/demand, enables mitigating power mismatches in the microgrid by power transfer from the main grid. The main grid could further provide reserve for the microgrid when the predicted variable generations are not materialized or when load forecast errors are high. However, the optimal microgrid scheduling and the UC problem in the main grid share a common objective, i.e. to determine the least cost operations of available resources to supply forecasted loads while taking prevailing operational constraints into consideration. Although sharing a common objective, the mentioned

The Economics of Microgrids, First Edition. Amin Khodaei and Ali Arabnya.
© 2024 The Institute of Electrical and Electronics Engineers, Inc.
Published 2024 by John Wiley & Sons, Inc.

differences would not allow a direct application of existing UC methods to the microgrid optimal scheduling problem. The rapid development of microgrids calls for new methodologies to comprehensively model all the active components in microgrids and particularly focus on microgrid islanding requirements when the main grid power is not available.

Microgrid optimal scheduling has been extensively investigated in the literature. The existing energy management system architecture for microgrids are reviewed in [4], where centralized and distributed models are identified as common microgrid control schemes. The centralized model collects all the required information for the microgrid scheduling and performs a centralized operation and control [5–9]. In the distributed model, however, each component is considered as an agent with the ability of discrete decision-making. The optimal schedule is obtained using iterative data transfers among agents [10–12]. Both control schemes offer benefits and drawbacks, but the centralized model is more desirable as it ensures a secure microgrid operation and is more suitable for application of optimization techniques. The main drawbacks of the centralized scheme are reduced flexibility in adding new components and extensive computational requirements [4].

Microgrid islanding studies are very limited in the literature. Reference [13] proposes an economic dispatch model for a microgrid that applies additional reserve constraints to enable islanding. Reference [14] presents a load management model to improve microgrid resilience following islanding, taking into account the microgrid's limited energy storage capability and frequency response. A method to determine the amount of storage required to meet reliability targets and guarantee island-capable operation with variable generation is proposed in [15]. In [16], storage systems are applied in microgrids to balance power, smooth out load, reduce power exchange with the main grid in the grid-connected mode, and ensure successful transition to the islanded mode.

This chapter presents a centralized microgrid optimal scheduling model, which considers multi-period islanding constraints. The objective is to minimize the day-ahead, grid-connected operations cost of the microgrid using available generation resources, energy storage systems, adjustable loads, and the main grid power, subject to prevailing operational constraints. The solution is examined for islanded operations to ensure the microgrid has sufficient online capacity for quickly switching to the islanded mode if required. An islanding criterion is proposed that demonstrates the resiliency of the microgrid to operate in islanded mode for a variety of time durations. An iterative model based on the Benders' decomposition is employed to couple grid-connected operations (as a master problem) and islanded operations (as a subproblem). The iterative model significantly reduces the problem's computation burdens and enables a quick solution. Problems are modeled using mixed-integer programming, which facilitates addition of new components to the microgrid.

The proposed model is developed specifically for microgrids. The proposed model effectively considers uncertain microgrid islanding (from islanding time and duration standpoints) in the microgrid optimal scheduling problem and enables the microgrid to operate in the islanded mode and adequately supply the local loads when the time and extent of the main grid disturbance are unknown. The islanding duration is considered via the novel criterion of multi-period islanding, which refers to the islanding event that takes several hours. The proposed model is comprehensive yet flexible in adding new components to the microgrid and benefits from a decomposed model that reduces computation burdens and makes it suitably applicable to centralized microgrid scheduling schemes.

2.2 Dynamics of Optimal Scheduling in Microgrids

Figure 2.1 depicts the flowchart of the microgrid optimal scheduling model. The problem is decomposed into a grid-connected operations master problem and an islanded operations subproblem. The master problem determines the optimal

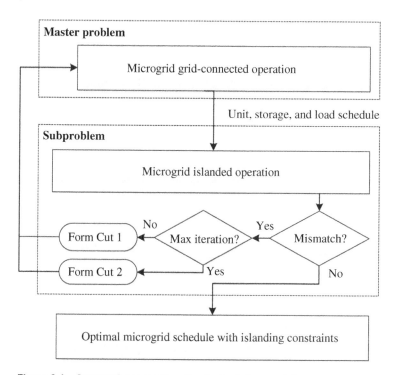

Figure 2.1 Proposed microgrid optimal scheduling model.

commitment and dispatch of available dispatchable units, charging and discharging schedules of energy storage systems, schedule of adjustable loads, and the power transfer with the main grid. The optimal schedule is used in the subproblem to examine the microgrid generation adequacy and confirm an uninterrupted supply of loads for a variety of islanding scenarios. If the islanding is not feasible, i.e. microgrid does not have sufficient online capacity to supply the local load, a Benders' cut, i.e. Cut 1, based on the UC and energy storage system schedules is generated and sent back to the master problem for revising the current solution. The Benders' cut indicates that power mismatches in the subproblem can be mitigated by readjusting the UC and energy storage system schedules in the master problem. The revised solution will be examined in the next iteration of the islanding subproblem. The iterative process continues until all islanding scenarios are feasible. It is possible, however, in some scenarios that changes in UC and energy storage system schedules do not provide required online capacity to guarantee a feasible islanding. In this situation, a secondary Benders' cut, i.e. Cut 2, is generated based on adjustable load schedules. This cut would revise the adjustable loads' specified operating time intervals to shift the load and accordingly enable the islanding. The inconvenience realized by consumers as a result of this change is penalized in the objective. This Benders' cut indicates that power mismatch in the subproblem can be mitigated by readjusting load schedules in addition to UC and energy storage system schedules in the master problem. The final solution is obtained when all islanding scenarios are guaranteed to be feasible. Note that Cuts 1 and 2 are represented in the form of inequality constraints, which provide a lower estimate of the total mismatch in the subproblem as a function of scheduling variables in the master problem [17].

Day-ahead schedules are calculated for the master problem and the subproblem, i.e. a 24-hour scheduling horizon is considered. Any other scheduling horizon can be selected based on the master controller's discretion without any change in the model. Selection of a 24-hour scheduling horizon, however, would enable microgrid master controller to benefit from day-ahead market price forecasts provided by the utility company and also keep track of energy storage system's daily charging/discharging cycles. The dispatchable units' commitments and energy storage systems charging/discharging schedules will be determined in the master problem and remain unchanged in the subproblem. The microgrid's fixed load and generation of non-dispatchable units are forecasted with acceptable accuracy. The market price at the PCC, i.e. the price at which microgrid purchases the main grid power and sells excess power to the main grid, is also forecasted. It is assumed that microgrid components are highly reliable and are not subject to outages during the scheduling horizon.

2.2.1 $T - \tau$ Islanding Criterion

The microgrid must be able to switch to islanded mode at any given time in response to disturbances in the main grid. The microgrid would be resynchronized with the main grid once the disturbance is removed. The microgrid master controller, however, is not aware of the disturbance time and duration. Therefore, microgrid resources are to be scheduled in a way that local loads are supplied with no interruption using only local resources, i.e. an islanded operation, for an unknown time extent.

To characterize the microgrid's capability in responding to time-varying islanding requirements, a $T - \tau$ islanding criterion is proposed. T denotes the number of hours in the scheduling horizon, and τ represents the number of consecutive hours that the microgrid can operate in the islanded mode. As an example, a $T - 2$ islanding criterion requires that the microgrid be able to operate in the islanded mode for any two-hour period once it is switched from grid-connected to the islanded mode. In the two successive islanding hours, the microgrid load is fully supplied from local resources since the power cannot be transferred from the main grid. This criterion represents an effective approach in ensuring microgrid resiliency and online generation adequacy in multi-hour islanding operations.

In addition to uncertainty in the microgrid islanding time and duration, forecast errors associated with the market price, the non-dispatchable unit generation, and loads add additional uncertainty to the microgrid optimal scheduling problem.

2.3 An Economic Model for Microgrid Optimal Scheduling with Multi-Period Islanding

2.3.1 Grid-Connected Operations

The objective of the grid-connected operation master problem is to minimize the microgrid's total operation cost as follows:

$$\text{Min} \sum_{t} \sum_{i \in G} [F_i(P_{it})I_{it} + SU_i + SD_i] + \sum_{t} \rho_t P_{M,t} \tag{2.1}$$

where $F(.)$ is the generation cost, SD is the shutdown cost, SU is the startup cost of each distributed energy resources (DER), P_M is the main grid power, and ρ is the market price. The first term in the objective is the operations cost of microgrid dispatchable units, which includes generation, startup and shutdown costs over the entire scheduling horizon. The generation cost is commonly represented by a quadratic function; however, it could be simply approximated by a piecewise linear model. The second term is the cost of power transfer from the main grid based on the market price at PCC. When the microgrid's excess power is sold back to the main grid, $P_{M,t}$ would be negative, thus this term would represent a benefit, rather than a cost, for the microgrid. The objective is subject to generating unit, energy storage system, and load constraints, as follows:

$$\sum_i P_{it} + P_{M,t} = \sum_d D_{dt} \qquad \forall t \tag{2.2}$$

$$-P_M^{\max} \leq P_{M,t} \leq P_M^{\max} \qquad \forall t \tag{2.3}$$

$$P_i^{\min} I_{it} \leq P_{it} \leq P_i^{\max} I_{it} \qquad \forall i \in G, \forall t \tag{2.4}$$

$$P_{it} - P_{i(t-1)} \leq UR_i \qquad \forall i \in G, \forall t \tag{2.5}$$

$$P_{i(t-1)} - P_{it} \leq DR_i \qquad \forall i \in G, \forall t \tag{2.6}$$

$$T_i^{\text{on}} \geq UT_i\left(I_{it} - I_{i(t-1)}\right) \qquad \forall i \in G, \forall t \tag{2.7}$$

$$T_i^{\text{off}} \geq DT_i\left(I_{i(t-1)} - I_{it}\right) \qquad \forall i \in G, \forall t \tag{2.8}$$

$$P_{it} \leq P_{it}^{\text{dch, max}} u_{it} - P_{it}^{\text{ch, min}} v_{it} \qquad \forall i \in S, \forall t \tag{2.9}$$

$$P_{it} \geq P_{it}^{\text{dch, min}} u_{it} - P_{it}^{\text{ch, max}} v_{it} \qquad \forall i \in S, \forall t \tag{2.10}$$

$$u_{it} + v_{it} \leq 1 \qquad \forall i \in S, \forall t \tag{2.11}$$

$$C_{it} = C_{i(t-1)} - P_{it} \qquad \forall i \in S, \forall t \tag{2.12}$$

$$0 \leq C_{it} \leq C_i^{\max} \qquad \forall i \in S, \forall t \tag{2.13}$$

$$T_i^{\text{ch}} \geq MC_i\left(u_{it} - u_{i(t-1)}\right) \qquad \forall i \in S, \forall t \tag{2.14}$$

$$T_i^{\text{dch}} \geq MD_i\left(v_{it} - v_{i(t-1)}\right) \qquad \forall i \in S, \forall t \tag{2.15}$$

$$D_{dt}^{\min} z_{dt} \leq D_{dt} \leq D_{dt}^{\max} z_{dt} \qquad \forall d \in D, \forall t \tag{2.16}$$

$$\sum_{t \in [\alpha_d, \beta_d]} D_{dt} = E_d \qquad \forall d \in D \tag{2.17}$$

$$T_d^{\text{on}} \geq MU_d\left(z_{dt} - z_{d(t-1)}\right) \qquad \forall i \in D, \forall t \tag{2.18}$$

where b is an index for energy storage systems, ch is the superscript for energy storage system charging mode, d is the index for loads, dch is the superscript for energy storage system discharging mode, i is the index for DERs, s is the index for scenarios, and t is the index for time. Sets D, G, and S represent adjustable loads, dispatchable units, and energy storage systems, respectively. Parameters of this model include DR for ramp-down rate, DT for minimum downtime, E for load total required energy, MC for minimum charging time, MD for minimum discharging time, MU for minimum operating time, UR for ramp-up rate, UT for minimum uptime, and α, β for specified start and end times of adjustable loads. The problem also includes multiple variables. C is the energy storage system state of charge (SOC), D is the load demand, I is the commitment state of the dispatchable unit, P is the DER output power, T^{ch} is the number of successive charging hours, T^{dch} is the number of successive discharging hours, T^{on} is the number of successive ON hours, T^{off} is the number of successive OFF hours, u is the energy storage system discharging state, v is the energy storage system charging state, and z is the adjustable load state.

The power balance Eq. (2.2) ensures that the sum of power generated by DERs (i.e. dispatchable and non-dispatchable units and energy storage systems) and the power from the main grid matches the hourly load. The forecasted generation of non-dispatchable units is used in (2.2), where it can be treated as a negative load. The power of energy storage systems can be positive (discharging), negative (charging), or zero (idle). The main grid power can be positive (import), negative (export), or zero. The power transfer with the main grid is limited by the flow limits of the line connecting microgrid to the main grid (2.3). The dispatchable unit generation is subject to minimum and maximum generation capacity limits (2.4), ramp-up and ramp-down rate limits (2.5) and (2.6), and minimum up and downtime limits (2.7) and (2.8). The UC state, I_{it}, is one when unit is committed and is zero otherwise. A dispatchable unit can further be subject to fuel and emission limits based on the unit type.

The energy storage system's power is subject to charging and discharging minimum and maximum limits depending on its mode (2.9) and (2.10). When charging, the charging state v_{it} is one and discharging state u_{it} is zero, hence minimum and maximum charging limits are imposed. Similarly, when discharging, the discharging state u_{it} is one and charging state v_{it} is zero; hence, minimum and maximum discharging limits are imposed. Since the energy storage system charging power is considered negative, the associated limits are denoted with a minus sign. Only one of the charging or discharging modes at every hour is possible (2.11). Energy storage system SOC is calculated based on the amount of charged/discharged power (2.12) and restricted with capacity limits (2.13). The SOC at $t = 1$ is calculated based on SOC at the last hour of the previous scheduling horizon. It is also assumed that energy storage systems maintain similar SOC at the beginning and end of the scheduling horizon. Energy storage systems are subject to minimum charging and discharging time limits, respectively (2.14) and (2.15), which are the minimum number of consecutive hours that energy storage systems should maintain charging/discharging once the operational mode is changed.

Adjustable loads are subject to minimum and maximum rated powers (2.16). When load is consuming power, the associated scheduling state z_{dt} would be one; it is zero otherwise. Each load consumes the required energy to complete an operating cycle in time intervals specified by consumers (2.17). α_d and β_d represent the start and end operating times of an adjustable load, respectively. Certain loads may be subject to minimum operating time, which is the number of consecutive hours that a load should consume power once it is switched on (2.18).

2.3.2 Islanded Operations

The objective of the islanded operation subproblem for an islanding scenario s is to minimize the power mismatches as in (2.19).

$$\text{Min } w_s = \sum_t (SL_{1,ts} + SL_{2,ts}) \tag{2.19}$$

$$\sum_i P_{its} + P_{M,ts} + SL_{1,s} - SL_{2,s} = \sum_d D_{dts} \quad \forall t \tag{2.20}$$

$$I_{its} = \hat{I}_{it} \quad \lambda_{its} \qquad \forall i \in G, \forall t \tag{2.21}$$

$$u_{its} = \hat{u}_{it} \quad \mu_{its}^{dch} \qquad \forall i \in S, \forall t \tag{2.22}$$

$$v_{its} = \hat{v}_{it} \quad \mu_{its}^{ch} \qquad \forall i \in S, \forall t \tag{2.23}$$

$$z_{dts} = \hat{z}_{dt} \quad \pi_{dts} \qquad \forall d \in D, \forall t \tag{2.24}$$

$$-P_M^{\max} U_{ts} \le P_{M,ts} \le P_M^{\max} U_{ts} \qquad \forall t \tag{2.25}$$

$$P_i^{\min} I_{its} \le P_{its} \le P_i^{\max} I_{its} \qquad \forall i \in G, \forall t \tag{2.26}$$

$$P_{its} - P_{i(t-1)s} \le UR_i \qquad \forall i \in G, \forall t \tag{2.27}$$

$$P_{i(t-1)s} - P_{its} \le DR_i \qquad \forall i \in G, \forall t \tag{2.28}$$

$$P_{its} \le P_{it}^{dch,\max} u_{its} - P_{it}^{ch,\min} v_{its} \qquad \forall i \in S, \forall t \tag{2.29}$$

$$P_{its} \ge P_{it}^{dch,\min} u_{its} - P_{it}^{ch,\max} v_{its} \qquad \forall i \in S, \forall t \tag{2.30}$$

$$C_{its} = C_{i(t-1)s} - P_{its} \qquad \forall i \in S, \forall t \tag{2.31}$$

$$0 \le C_{its} \le C_i^{\max} \qquad \forall i \in S, \forall t \tag{2.32}$$

$$D_{dt}^{\min} z_{dts} \le D_{dt} \le D_{dt}^{\max} z_{dts} \qquad \forall d \in D, \forall t \tag{2.33}$$

$$\sum_{t \in [\alpha_d, \beta_d]} D_{dts} = E_d \qquad \forall d \in D \tag{2.34}$$

where SL_1, SL_2 are slack variables, U is the outage state of the main grid line/ islanding state, and w is the power mismatch. Power balance Eq. (2.20) encompasses slack variables SL_1 and SL_2, which act as virtual generation and virtual load, respectively. Nonzero values for these variables denote a power mismatch in the microgrid. UCs, energy storage charging/discharging schedules, and load schedules are obtained from the grid-connected operation master problem. These given variables are replaced with local variables for each scenario to obtain associated dual variables (2.21)–(2.24). Dual variables are later used in this section to generate islanding cuts.

Main grid power transfer constraint is revised by including a binary outage state, i.e. U_{ts}. When the outage state is set to zero, the main grid power will be zero, and therefore, the microgrid is imposed to operate in the islanded mode. Islanding scenarios are generated using the outage state. In each scenario, the outage state will obtain 0–1 values based on the islanding duration and will be considered in the islanded operation subproblem as an input. The islanded operation subproblem is further subject to dispatchable unit generation and ramp rate limits (2.26)–(2.28), energy storage system power and capacity limits (2.29)–(2.32), and adjustable load power and energy limits (2.33) and (2.34).

A zero mismatch for the islanded operation subproblem ensures that the microgrid has sufficient committed generation and energy storage to independently supply the local load; hence it could switch to the islanded mode without interruption in the load supply. When the objective is not zero, however, islanding Cut 1 (2.35) is generated and added to the next iteration of the grid-connected operation master problem to revise the current microgrid schedule.

$$\hat{w}_s + \sum_{i \in G} \lambda_{its}(I_{it} - I_{its}) + \sum_{i \in S} \mu_{its}^{\text{dch}}(u_{it} - u_{its}) + \sum_{i \in S} \mu_{its}^{\text{ch}}(v_{it} - v_{its}) \leq 0, \qquad (2.35)$$

where λ_{its}, μ_{its}^{dch}, and μ_{its}^{ch} are dual variables of (2.21), (2.22), and (2.23), respectively. The islanding Cut 1 indicates that islanding mismatches can be mitigated by readjusting the microgrid schedule in the grid-connected operation master problem. Dual variables in the islanding cut are the incremental reduction in the objective function of the islanded operation subproblem. This cut results in a change in UC and energy storage system schedules based on islanding considerations. The iterative process continues until power mismatches in all islanding scenarios reach zero. However, it is probable that after a certain number of iterations the islanding is not guaranteed, i.e. by revising UC and energy storage system schedules, a zero mismatch in all islanding scenarios is not obtained. To resolve this issue the schedule of adjustable loads would be revised using the following cut, i.e. Cut 2:

$$\begin{aligned} \hat{w}_s + \sum_{i \in G} \lambda_{its}(I_{it} - I_{its}) + \sum_{i \in S} \mu_{its}^{\text{dch}}(u_{it} - u_{its}) + \sum_{i \in S} \mu_{its}^{\text{ch}}(v_{it} - v_{its}) \\ + \sum_{i \in D} \pi_{dts}(z_{dt} - z_{dts}) \leq 0, \end{aligned} \qquad (2.36)$$

where π_{dts} is the dual variable of (2.24). Cut 2 enables a simultaneous change in UC, energy storage system schedules, and adjustable load schedules to guarantee a feasible islanding. To change the adjustable load schedule its specified start and end operating times are revised, in which the new operating time interval is represented by $\left[\alpha_d^{new}, \beta_d^{new}\right]$. The inconvenience for consumers due to the change in operating time interval is modeled with a penalty term (2.37) and added to the objective (2.1).

$$\sum_{d \in D} K_d \Delta_d, \tag{2.37}$$

where K_d is the inconvenience penalty factor and Δ_d is the deviation in adjustable load operating time interval. Additional constraints (2.38)–(2.40) are added to the grid-connected operation master problem to reflect this change. Equation (2.38) measures the total deviation in the operating time interval from original specified values, and (2.39) and (2.40) ensure that the new time interval spans a wider time range than the original one.

$$\Delta_d = \left(\beta_d^{new} - \alpha_d^{new}\right) - \left(\beta_d - \alpha_d\right) \qquad \forall i \in D \tag{2.38}$$

$$\beta_d^{new} \geq \beta_d \qquad \forall i \in D \tag{2.39}$$

$$\alpha_d^{new} \leq \alpha_d \qquad \forall i \in D \tag{2.40}$$

The consumer inconvenience is penalized with a constant penalty factor K_d. This penalty factor could be used to prioritize the loads with regard to sensitivity when operating within the specified time intervals. A higher value for K_d represents a less flexible load in terms of operating time, which gains a lower priority for time interval adjustment. The value for K_d should be selected reasonably higher than the generation cost of units and the market price; therefore, the grid-connected operation master problem would consider the change in load operating time intervals as a last resort.

A period of one hour is considered for modeling the master problem and the subproblem. Accordingly, islanding duration is considered as an integer multiple of one hour. Shorter time periods, however, could be considered without significant change in the proposed model. The selection of a proper time period for scheduling represents a tradeoff between the solution accuracy and the computation time. Shorter time periods would embrace more data and provide more accurate solutions while increasing computation requirements.

2.4 Case Study

A microgrid with four dispatchable units, two non-dispatchable units, one energy storage system, and five adjustable loads is used to analyze the proposed microgrid optimal scheduling model. The problem is implemented on a 2.4-GHz personal computer using CPLEX 11.0 [18]. The characteristics of units, energy storage systems, and adjustable loads are given in Tables 2.1–2.3, respectively. The forecasted values for microgrid hourly fixed load, non-dispatchable units' generation, and market price over the 24-h horizon are given in Tables 2.4–2.6, respectively.

Table 2.1 Characteristics of generating units.

Unit	Type	Cost coefficient ($/MWh)	Min–max capacity (MW)	Min up/down time (h)	Ramp up/down rate (MW/h)
G1	D	27.7	1–5	3	2.5
G2	D	39.1	1–5	3	2.5
G3	D	61.3	0.8–3	1	3
G4	D	65.6	0.8–3	1	3
G5	ND	0	0–1	—	—
G6	ND	0	0–1.5	—	—

D, dispatchable; ND, non-dispatchable.

Table 2.2 Characteristics of the energy storage system.

Storage	Capacity (MWh)	Min–max charging/discharging power (MW)	Min charging/discharging time (h)
ESS	10	0.4–2	5

Table 2.3 Characteristics of adjustable loads.

Load	Type	Min–max capacity (MW)	Required energy (MWh)	Initial start-end time (h)	Min up time (h)
L1	S	0–0.4	1.6	11–15	1
L2	S	0–0.4	1.6	15–19	1
L3	S	0.02–0.8	2.4	16–18	1
L4	S	0.02–0.8	2.4	14–22	1
L5	C	1.8–2	47	1–24	24

S, shiftable; C, curtailable.

Table 2.4 Microgrid hourly fixed load.

Time (h)	1	2	3	4	5	6
Load (MW)	8.73	8.54	8.47	9.03	8.79	8.81
Time (h)	7	8	9	10	11	12
Load (MW)	10.12	10.93	11.19	11.78	12.08	12.13
Time (h)	13	14	15	16	17	18
Load (MW)	13.92	15.27	15.36	15.69	16.13	16.14
Time (h)	19	20	21	22	23	24
Load (MW)	15.56	15.51	14.00	13.03	9.82	9.45

Table 2.5 Generation of non-dispatchable units.

Time (h)	1	2	3	4	5	6
G5	0	0	0	0	0.63	0.80
G6	0	0	0	0	0	0
Time (h)	7	8	9	10	11	12
G5	0.62	0.71	0.68	0.35	0.62	0.36
G6	0	0	0	0	0	0.75
Time (h)	13	14	15	16	17	18
G5	0.4	0.37	0	0	0.05	0.04
G6	0.81	1.20	1.23	1.28	1.00	0.78
Time (h)	19	20	21	22	23	24
G5	0	0	0.57	0.60	0	0
G6	0.71	0.92	0	0	0	0

Table 2.6 Hourly market price.

Time (h)	1	2	3	4	5	6
Price ($/MWh)	15.03	10.97	13.51	15.36	18.51	21.8
Time (h)	7	8	9	10	11	12
Price ($/MWh)	17.3	22.83	21.84	27.09	37.06	68.95
Time (h)	13	14	15	16	17	18
Price ($/MWh)	65.79	66.57	65.44	79.79	115.45	110.28
Time (h)	19	20	21	22	23	24
Price ($/MWh)	96.05	90.53	77.38	70.95	59.42	56.68

The following cases are studied:

Case 0: Grid-connected microgrid optimal scheduling;
Case 1: Optimal scheduling with $T-1$ islanding criterion;
Case 2: Optimal scheduling with $T-2$ islanding criterion;
Case 3: Sensitivity with regard to market price forecast errors; and
Case 4: Sensitivity with regard to the problem size.

Case 0: The grid-connected microgrid optimal scheduling is studied for a 24-hour horizon. The DER schedule, including dispatchable UC states and the energy storage system schedule, is shown in Table 2.7. The commitment state is 1 when the unit is on and is zero otherwise. The energy storage system's charging, discharging, and idle states are represented by -1, 1, and 0, respectively. The economic unit 1 is committed to the entire scheduling horizon as it offers low-cost power. Units 2–4 are committed and dispatched at the maximum capacity when the market price exceeds the cost coefficient of these units. The energy storage system is charged at low price hours 1–6 and discharged at high price hours 16–20, shifting a total load of 10 MWh from peak hours to off-peak hours. Adjustable loads are scheduled to minimize the consumption cost and adopt to start and end times provided by the consumers, as shown in Table 2.8. The microgrid grid-connected operation cost is \$11,183. The result indicates that the microgrid would decide on the supply source only based on economic considerations. A unit is committed only when its cost coefficient is lower than the market price. It would accordingly generate its maximum power to sell the excess power to the main grid and increase microgrid savings (i.e. to further reduce the total operation cost). The microgrid would also discharge the energy storage system at peak hours, when the market price is at its highest, for the same economic reasons.

Case 1: The microgrid optimal scheduling is studied considering a $T-1$ islanding criterion. Twenty-four scenarios are considered, each including a one-hour islanding. The $T-1$ islanding is imposed as a robust requirement, i.e. all single-hour islanding scenarios must be satisfied without causing load curtailment. In the islanded mode, the power transfer from the main grid is zero; therefore, sufficient capacity is committed in the grid-connected mode to enable a quick switch to the islanded mode without interruption in load supply. The optimal and feasible solution is obtained in three iterations with an execution time of six seconds. The master problem solution in iteration 1 is similar to the solution in Case 0 (as shown in Tables 2.7 and 2.8). However, this schedule results in a total mismatch of 75.63 MWh in islanding scenarios. Cut 1 is generated based on the mismatch in each islanding scenario for revising the obtained dispatchable UC and the energy storage system schedule in the master problem. With the revised schedule, the second iteration's total mismatch is reduced to 3.22 MWh. Since the mismatch is not zero, another Cut 1 is generated and sent back to the master problem to further

Table 2.7 DER schedule in Case 0.

	Hours (1–24)																							
G1	1	1	1	1	1	1	1	1	1	1	1	1	1	1	1	1	1	1	1	1	1	1	1	1
G2	0	0	0	0	0	0	0	1	1	1	1	1	1	1	1	1	1	1	1	1	1	1	1	1
G3	0	0	0	0	0	0	0	0	0	1	1	1	1	1	1	1	1	1	1	1	1	1	0	0
G4	0	0	0	0	0	0	0	0	0	0	1	1	0	1	1	1	1	1	1	1	1	0	0	0
ESS	−1	−1	−1	−1	−1	0	0	0	0	0	0	0	0	1	1	1	1	0	0	0	0	0	0	0

Table 2.8 Adjustable load schedule in Case 0.

	Hours (1–24)																							
L1	0	0	0	0	0	0	0	0	0	0	0	1	1	1	0	0	0	0	0	0	0	0	0	0
L2	0	0	0	0	0	0	0	0	0	0	0	1	1	0	0	0	0	0	0	0	0	0	0	0
L3	0	0	0	0	0	0	0	0	0	0	0	0	0	1	1	0	1	0	0	0	0	0	0	0
L4	0	0	0	0	0	0	0	0	0	0	1	1	1	0	0	0	0	0	1	0	1	1	0	0
L5	1	1	1	1	1	1	1	1	1	1	1	1	1	1	1	1	1	1	1	1	1	1	1	1

revise the schedule. The third iteration mismatch reaches a value of zero, which means that microgrid islanding criterion is satisfied and islanding is feasible in all scenarios. The feasible islanding solution is obtained using only Cut 1, so there would be no need to form Cut 2 and change the optimal schedule of adjustable loads.

Table 2.9 shows the commitment results considering $T - 1$ islanding, where the bold values highlight changes in the solution from Case 0. Additional units are committed each hour to ensure an uninterrupted supply of loads when the microgrid is islanded. These units are dispatched at their minimum capacity as their generation is not economical. The energy storage system is discharged at a slower rate, i.e. seven hours compared to five hours in Case 0, to cooperate in the microgrid islanding when the available unit capacity cannot completely supply the local load. The adjustable load schedule remains unchanged, since Cut 2 is not formed, and the purchased energy from the main grid is reduced by 11.13 MWh, which is due to utilization of additional local resources. The microgrid operation cost is $11,674.55. The operation cost difference between Cases 0 and 1, i.e. $491.55, is considered as the cost of islanding.

The result indicates that when considering microgrid islanding, additional units, which do not necessarily offer economic merits, have to be committed and maintained online. Although the microgrid schedule is significantly changed due to additional commitments, the total increase in the microgrid operation cost is only 4.4% compared to the grid-connected operation cost in Case 0. This small cost increase provides a huge benefit as microgrid islanding without load interruption is ensured.

Figure 2.2 depicts the main grid power transfer in Cases 0 and 1. In low-price hours, the power is purchased from the main grid as much as possible, i.e. 10 MW, equal to the line limit. The power purchase is reduced as the market price increases and generation of local resources becomes more economic. The sudden increase in hour 15 is due to the fact that the market price becomes lower than cost coefficient of dispatchable unit 4, and therefore this unit is turned off and the required power is purchased from the main grid. The main grid power transfer is almost similar in Cases 0 and 1. The minor differences in power transfer from the main grid in these two cases are a result of dispatchable units generating at their minimum capacity to enable a feasible islanding. The minimum generation of these units reduces the required power to be purchased from the main grid.

The role of the energy storage system is further investigated in this case by adding a second energy storage system with a capacity of 10 MWh, minimum-maximum rated power of 0.5–2.5 MW, and minimum charging/discharging time of four hours. Figure 2.3 compares the charging/discharging schedule of these two energy storage systems. Both energy storage systems are charged at low-price hours. Energy storage system 2, with a higher charging rate, is charged at lower

Table 2.9 DER schedule in Case 1.

	Hours (1–24)																							
G1	1	1	1	1	1	1	1	1	1	1	1	1	1	1	1	1	1	1	1	1	1	1	1	1
G2	1	1	1	1	1	1	1	1	1	1	1	1	1	1	1	1	1	1	1	1	1	1	1	1
G3	1	1	1	0	1	1	1	1	1	1	1	1	1	1	1	1	1	1	1	1	1	1	1	1
G4	0	0	0	0	0	0	0	0	0	0	1	1	1	1	1	1	1	1	1	1	1	1	0	0
ESS	−1	−1	−1	−1	0	0	0	0	0	0	0	0	0	0	0	0	0	0	1	1	1	0	0	0

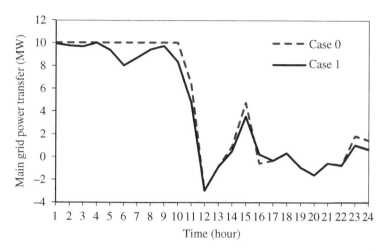

Figure 2.2 Main grid power transfer in Cases 0 and 1.

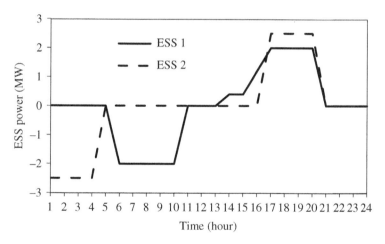

Figure 2.3 Charging/discharging schedules of energy storage systems.

price hours, while charging of energy storage system 1 is delayed by five hours. An overlap between charging schedules requires additional generation of dispatchable units; however, these units are not economical at these hours and are dispatched at their minimum power output. Therefore, charging of energy storage system 1 is delayed until supplied by the main grid power. Both energy storage systems are discharged during peak hours when the market price is high. Energy storage system 1 is discharged over an extended period of time to facilitate a feasible islanding.

Case 2: The microgrid optimal scheduling is studied considering a $T - 2$ islanding criterion, where the microgrid should have the islanding capability for every consecutive two-hour interruption in the main grid power. Initial grid-connected schedule results in a total mismatch of 145.73 and 6.45 MWh in the first two iterations of the subproblem. The generated cut based on UCs and the energy storage system schedule, i.e. Cut 1, does not further reduce the mismatch, hence cannot ensure a feasible islanding at hours 17 and 18. When iteration has reached its maximum limit, here 10, the subproblem generates Cut 2, which includes the schedule of adjustable loads, and sends it to the master problem. Cut 2 is formed to reduce the mismatch, i.e. 6.45 MWh, by revising the schedule of adjustable loads in addition to dispatchable units and energy storage systems. A penalty factor is added to the master problem to minimize the change in the operating time interval of adjustable loads. The penalty cost is assumed to be $100 for every hour of deviation from the specified start and end times. The microgrid operation cost is reduced to $11,657.07; however, an inconvenience cost of $40 (=100 × 0.4) is added to the total microgrid operation cost. 0.4 MW of load 2 is scheduled for hour 14, which is outside specified time interval by the consumer. Dispatchable unit and energy storage system schedules remain unchanged compared to Case 1 (see Table 2.9). The solution is obtained in 24 seconds. The result obtained in this case illustrates that to enable the microgrid islanding not only the UC and storage schedules but in some cases adjustable load schedules should be revised. If the microgrid cannot change the operating time interval of adjustable loads, it would have to inevitably curtail the load when in islanded mode to match the reduced load with available generation. This action would be more undesirable for consumers than revising the load operating time intervals.

The microgrid optimal scheduling problem is further solved for a variety of islanding criteria, from $T - 1$ to $T - 12$. Figure 2.4 shows the results and illustrates that a larger number of islanding hours would increase the microgrid operation cost. This increase is a direct result of inconvenience recognized by consumers, which is added to the objective as a cost. It is possible that a revised load schedule provides a feasible solution for a variety of islanding criteria, as it happened at $\tau = 2$ to $\tau = 5$ and also $\tau = 8$ to $\tau = 10$. Furthermore, the cost difference among islanding hours is very small compared to the grid-connected operation (shown at point zero in Figure 2.4). It demonstrates that the major cost of islanding occurs at $T - 1$ islanding. Additional islanding hours, however, could be performed at a small expense.

Case 3: A sensitivity analysis is performed to study the impact of forecast errors on microgrid optimal scheduling solutions. Thousand scenarios are generated to simulate market price forecast errors based on a uniform random error of ±30% of the hourly forecasted price in Table 2.6. The microgrid's optimal scheduling with $T - 1$ islanding is performed for all scenarios. The microgrid operation cost of 1000

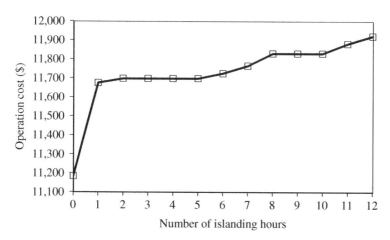

Figure 2.4 Total operation cost as a function of number of islanding hours.

scenarios falls within small lower and upper bounds of [$11,433.89, $11,759.85], which corresponds to a deviation of [−2.06%, 0.73%] from the solution in Case 1. This study shows that even with large forecast errors, acceptable solutions can be obtained using the proposed model. Similarly, 1000 scenarios are considered for load forecast error of ±10% of the hourly forecasted load. Obtained solutions deviate from the solution in Case 1 within [−3.95%, 2.96%]. Although the load forecast error is much lower than the price forecast error, deviation from the solution in Case 1 is much higher. This result suggests that the microgrid operation cost is highly sensitive to load forecast errors, as small errors may translate into huge changes in the operation cost. It is worth mentioning that in the proposed method fixed and adjustable loads are modeled separately. The main source of error in load forecasts is the unpredictable schedules of adjustable loads that depend on market prices and consumer preferences. The fixed load, on the other hand, can be forecasted with an acceptable level of accuracy in the short-term operation of the microgrid.

Performing a similar study for a ±30% forecast error in non-dispatchable generation, a deviation of [−1.75%, 0.96%] from solution in Case 1 is obtained. The low impact of non-dispatchable generation forecasts on the microgrid's optimal operation is due to the fact that non-dispatchable generation represents a small portion of the total generation in the microgrid, which is less than 14% of the total installed capacity in the studied microgrid. Therefore, even a large change in the generation of these resources would not change the results significantly. The small ratio of non-dispatchable units' capacity compared to dispatchable units' capacity in a microgrid is due to the fact that microgrid master controller should rely on dispatchable units for a feasible islanding in case the forecasted non-dispatchable generation is not materialized. Furthermore, non-dispatchable units offer a variable generation, i.e. the power output is not always available and may reach zero generation for several hours during the scheduling horizon. Thus, the energy produced by these resources could be much lower than the generation of a dispatchable unit of the same size.

Case 4: To demonstrate the effectiveness of the proposed model in solving the microgrid optimal scheduling problem in a reasonable amount of time, the problem is solved for a variety number of adjustable loads. The number of adjustable loads is changed from 10 to 100 instead of considering only five aggregated adjustable loads as in previous cases. The microgrid optimal scheduling problem with $T-1$ islanding criterion is solved using both integrated and decomposed models. Figure 2.5 compares the computation time between these two models. Using the proposed model, when the number of loads is increased, the computation time increases almost linearly. The computation time for 100 adjustable loads is about 10 times the computation time when 10 adjustable loads are considered. Using the integrated model, by increasing the number of adjustable loads, the computation time increases exponentially, whereas for 100 adjustable loads, the computation time is greater than 100 minutes.

The proposed model decomposes the problem into a master problem and a subproblem. All binary variables associated with the dispatchable units, energy storage systems, and adjustable loads are determined in the master problem, while the subproblem deals with linear variables and examines linear constraints. Furthermore, the islanding scenarios can be solved separately in the subproblem as there

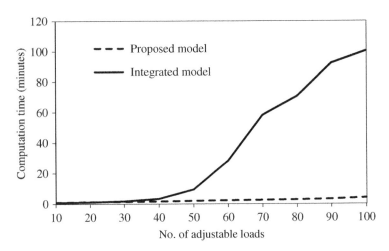

Figure 2.5 Comparison of computation time in integrated and decomposed models.

is no coupling constraint among islanding scenarios. Therefore, instead of solving a large-scale problem, several smaller problems are solved in an iterative manner, which would significantly reduce computation time.

2.5 Summary

In this chapter, a model for optimal microgrid operations scheduling considering multi-period islanding constraints is introduced. A novel islanding criterion was proposed for ensuring the generation adequacy of the microgrid in the islanded mode operations when the unpredicted disconnection from the main grid lasts more than one hour. The proposed criterion reflected the uncertainty in the duration of the main grid disturbance. Benders' decomposition method was employed to decouple the grid-connected operation and islanded operation problems. Islanding cuts were further utilized to couple these two problems. Mixed-integer programming was used to model microgrid components, which included loads, generating units, and energy storage systems. The proposed model was analyzed through numerical simulations, where it was shown that the islanding criterion would provide significant reliability benefits while slightly increasing the microgrid's total operations cost.

References

1 C. Hou, X. Hu, and D. Hui, "Hierarchical control techniques applied in micro-grid," in *IEEE Conf. on Power Syst. Tech. (POWERCON)*, Hangzhou, Oct. 2010.

2 A. G. Tsikalakis and N. D. Hatziargyriou, "Centralized control for optimizing microgrids," *IEEE Trans. Energy Convers.*, vol. 23, no. 1, pp. 241–248, 2008.

3 Federal Energy Management Program, "Using Distributed Energy Resources, a How-to Guide for Federal Facility Managers," DOE/GO-102002-1520, US Department of Energy, 2002.

4 D. E. Olivares, C. A. Canizares, and M. Kazerani, "A centralized optimal energy management system for microgrids," in *Power Energy Soc. Gen. Meeting*, Detroit, MI, July 2011.

5 A. G. Tsikalakis and N. D. Hatziargyriou, "Centralized control for optimizing microgrids operation," *IEEE Trans. Energy Convers.*, vol. 23, no. 1, pp. 241–248, 2008.

6 N. Hatziargyriou, G. Contaxis, M. Matos, J. A. P. Lopes, G. Kariniotakis, D. Mayer, J. Halliday, G. Dutton, P. Dokopoulos, A. Bakirtzis, J. Stefanakis, A. Gigantidou, P. O'Donnell, D. McCoy, M. J. Fernandes, J. M. S. Cotrim, and A. P. Figueira, "Energy management and control of island power systems with increased penetration from

renewable sources," in *IEEE-PES Winter Meeting*, New York, NY, vol. 1, pp. 335–339, Jan. 2002.

7 M. Korpas and A. T. Holen, "Operation planning of hydrogen storage connected to wind power operating in a power market," *IEEE Trans. Energy Convers.*, vol. 21, no. 3, pp. 742–749, 2006.

8 S. Chakraborty and M. G. Simoes, "PV-microgrid operational cost minimization by neural forecasting and heuristic optimization," in *IEEE Industry Applications Society Annual Meeting*, Edmonton, AB, Canada, Oct. 2008.

9 H. Vahedi, R. Noroozian, and S. H. Hosseini, "Optimal management of microgrid using differential evolution approach," in *7th International Conference on the European Energy Market*, Madrid, Spain, 2010.

10 N. D. Hatziargyriou, A. Dimeas, A.G. Tsikalakis, J. A. P. Lopes, G. Karniotakis, and J. Oyarzabal, "Management of microgrids in market environment," in *International Conference on Future Power Systems*, Amsterdam, Netherlands, Nov. 2005.

11 T. Logenthiran, D. Srinivasan, and D. Wong, "Multi-agent coordination for DER in microgrid," in *IEEE International Conference on Sustainable Energy Technologies (ICSET)*, Singapore, Nov. 2008.

12 J. Oyarzabal, J. Jimeno, J. Ruela, A. Engler, and C. Hardt, "Agent based microgrid management system," in *International Conference on Future Power Systems*, Amsterdam, Netherlands, Nov. 2005.

13 A. Seon-Ju and M. Seung-Il, "Economic scheduling of distributed generators in a microgrid considering various constraints," in *IEEE Power & Energy Society General Meeting*, Calgary, AB, Canada, July 2009.

14 C. Gouveia, J. Moreira, C. L. Moreira, and J. A. Pecas Lopes, "Coordinating storage and demand response for microgrid emergency operation," *IEEE Trans. Smart Grid*, 4, 1898–1908, 2013.

15 J. Mitra, M. R. Vallem, "Determination of storage required to meet reliability guarantees on island-capable microgrids with intermittent sources," *IEEE Trans. Power Syst.*, vol. 27, no. 4, pp. 2360–2367, Nov. 2012.

16 R. Pawelek, I. Wasiak, P. Gburczyk, and R. Mienski, "Study on operation of energy storage in electrical power microgrid—modeling and simulation," in *International Conference on Harmonics and Quality of Power (ICHQP)*, Bergamo, Italy, Sept. 2010.

17 A. Conejo, E. Castillo, R. Minguez, and R. Garcia-Bertrand, "Decomposition Techniques in Mathematical Programming," New York: Springer, 2006.

18 ILOG CPLEX, ILOG CPLEX Homepage 2009 [Online]. Available: www.ilog.com, 2009.

3

Resilience Economics in Microgrids

3.1 The Art of Resilience

Significant impacts of weather-related incidents and natural disasters on electric power systems and subsequent economic and social disruptions have resulted in a growing global need in addressing the issue of power system resiliency. Resiliency represents the ability of power systems to withstand low-probability, high-impact incidents in an efficient manner while ensuring the least possible interruption in the supply of electricity and further enabling a quick recovery and restoration to the normal operation state. A novel and viable solution to resiliency issues in power systems is to deploy microgrids [1–6]. Microgrid deployment is becoming an increasingly attractive solution for electricity customers who cannot rely on the supply of power from the main grid, and/or are seeking economic benefits from locally generated power. Electricity customers within a microgrid could benefit from the power supplied from local resources when there is a failure in the main grid and the supply of power is interrupted. Furthermore, the microgrid's excess generation could be sold back to the main grid to provide financial benefits, primarily in terms of electricity payment reductions, for customers. Microgrids introduce unique resilience opportunities in power system operation and planning [7] by lowering the possibility of load shedding. Based on the definition, microgrids can connect and disconnect to/from the main grid distribution network and operate in either grid-connected or islanded modes. Microgrid islanding rapidly disconnects the microgrid from the main distribution network in order to protect microgrid components from upstream disturbances, and to shield voltage-sensitive loads from significant voltage drops in the main grid [8, 9]. Therefore, the microgrid islanded operation could provide an efficient solution for supplying local loads when the main grid power is not available, or the distribution network is faulty. The microgrid islanding capability represents this technology as a viable solution to address power system resiliency issues and has attracted significant

The Economics of Microgrids, First Edition. Amin Khodaei and Ali Arabnya.
© 2024 The Institute of Electrical and Electronics Engineers, Inc.
Published 2024 by John Wiley & Sons, Inc.

attention in recent years [10, 11]. Resiliency improvement is considered one of the complementary value propositions provided by microgrids, achieved via promoting the dispersion of power resources and islanding [12].

The resiliency benefits of microgrids are widely discussed in the literature; however, the mathematical modeling of microgrid optimal scheduling based on resiliency considerations is limited. Existing studies on microgrid resiliency can be found in [13–21]. In [13], adequacy constraints are considered to ensure sufficient operating margin in the microgrid's economic operation and cover critical loads in case of upstream network faults. The concept of intelligent distributed autonomous power systems is proposed in [14] for building a resilient and environment-friendly customer-based microgrid, where demand-side management is employed to ensure that critical loads are served during emergency conditions. A frequency droop control system for a microgrid is proposed in [15] to extend the capabilities of a resilient microgrid to a conventional distribution network. The study in [16] derives a sequence of control actions to be adopted for multi-microgrid system service restoration and subsequent operation in islanded mode. It is shown that the feasibility of the sequence of control actions allows the reduction of load restoration times and improves system resiliency. The study in [17] reports on recent research directed toward employing distributed multi-agent architectures to achieve resilient self-healing power systems through independent management of microgrids. It is further discussed that interconnected microgrids are viable solutions to power system resiliency issues. A microgrid to serve a residential area located in a hurricane path is proposed in [18]. Phase droop control and a central power management controller are proposed as control means to stabilize the system when it is subject to disturbances. In [19], the development of advanced microgrid load management functionalities to manage microgrid storage, electric vehicles, and load responsiveness, to improve microgrid resilience following the islanding events is presented. A planning approach for building resilient microgrids, by strategically deploying distributed generators in a distribution system, is proposed in [20], which aims at optimizing microgrid vulnerability, reliability, and economy. The optimization model is solved by a combination of multi-agent systems and particle swarm optimization. The study in [21] proposes multi-objective optimization for evaluating the sustainable design and operation of DERs in microgrids. A resiliency index is defined to account for the capacity of the power system to self-recover to a new normal state after experiencing an unanticipated catastrophic event.

In this chapter, a resiliency-oriented microgrid optimal scheduling model is proposed. A centralized scheduling model is adopted in which the master controller collects all the required information for microgrid scheduling and performs a centralized operation and control. The proposed centralized model ensures secure microgrid operation and is suitable for the application of optimization techniques. The microgrid's normal operation, when connected to the grid, is coordinated with a resilient operation for enabling a rapid switching between these two modes without interruption in the supply of loads. During normal operation, the microgrid is connected to the main grid distribution network; thus, it schedules local resources and transfers power with the main grid to minimize the microgrid's operation cost. In case of main grid disturbances, however, the microgrid is switched over to resilient operation, i.e. the islanded mode, to supply local loads and ensure a resilient operation. Prevailing uncertainties make the problem very challenging to solve. It is assumed that the microgrid operator would be able to estimate the time that the electricity infrastructure would be affected, i.e. the main grid supply would be interrupted, and accordingly, decide on the time that the microgrid must be switched to the islanded mode. However, the duration of the islanding is uncertain and depends on how fast the main grid would be repaired and restored. Figure 3.1 shows three stages of microgrid operation, which include before the incident (normal operation), during the main grid supply interruption (resilient operation), and after the main grid repair and restoration (resynchronized). Other uncertain factors include the microgrid load and variable renewable generation forecasts. Although these forecasts are performed for a short time period, i.e. from a few hours to a few days depending on the main grid repair and restoration time, forecast errors would significantly impact the microgrid optimal scheduling solution. A robust optimization method is employed to account for uncertainties in load and generation forecasts. The Benders' decomposition is employed to decouple and coordinate the normal operation and the resilient operation problems. The microgrid optimal scheduling model proposed in [22] is considered as the basis of this model and extended considerably to make it applicable for resiliency applications. In the proposed model, uncertainty in load, renewable generation, and time and duration of incidents are efficiently captured. The curtailment of local loads, when sufficient generation is not available, is also properly considered in the model for

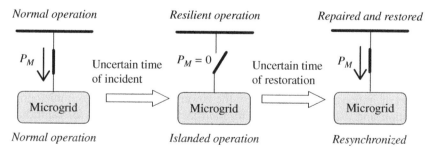

Figure 3.1 Microgrid operation before, during, and after the main grid disruption.

enhancing model practicality. In addition, dispatchable units' capability to revise their generation when switching to resilient operation is restricted via permissible power adjustment constraints.

3.2 The Impact of Uncertainty

Accurate modeling of microgrid components, as well as identification of sources of uncertainty, is required to ensure efficient microgrid optimal scheduling with resiliency considerations. Microgrid components, including fixed and adjustable loads, dispatchable and non-dispatchable units, and energy storage systems, are identified and discussed in detail in the literature [22]. The issue of uncertainty in microgrid scheduling, however, requires more investigation. Uncertainty refers to the fact that some factors, having a major influence on scheduling decisions, are not under the control of the microgrid master controller or cannot be predicted with certainty. Based on this definition, two major sources of uncertainty can be identified in the microgrid optimal scheduling problem, as follows: (i) forecast errors and (ii) main grid supply interruptions. The microgrid load, the non-dispatchable unit generation, and the market price cannot be accurately forecasted. Forecasts depend on a variety of factors that are out of control of the microgrid master controller, such as weather and site conditions, decisions of market players, and transmission network congestion, thus the forecast would not be completely accurate. This issue persists even in scheduling problems with relatively short horizons. Main grid supply interruptions are also uncertain as the time of incidents is unknown. Furthermore, depending on the range and severity of outages in the main grid, the required time to repair the power system and restore the power supply would vary. To ensure resiliency, the microgrid master controller must plan ahead for main grid supply interruptions while taking forecast uncertainties into account and accordingly perform seamless islanding when required.

Figure 3.2 illustrates the proposed resiliency-oriented microgrid optimal scheduling model. The problem is decomposed into a normal operation problem and a resilient operation problem. The normal operation problem determines the optimal schedule of units, energy storage, and adjustable loads, as well as the power transfer with the main grid. The optimal schedule is tested in the resilient operation problem to ensure generation adequacy for feasible islanding. The resilient operation problem minimizes the power mismatches between microgrid generation and load. A robust optimization method is employed for capturing uncertainties, in which it is assumed that uncertain parameters belong to convex and bounded uncertainty intervals. Forecast uncertainties are captured by determining the solution for the worst-case scenario in the resilient operation problem, i.e. the highest mismatch that would result when uncertain parameters fluctuate

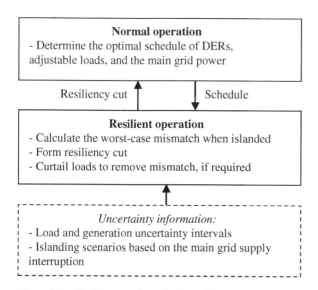

Figure 3.2 Resiliency-oriented microgrid optimal scheduling model.

in their associated uncertainty intervals. The uncertainty of the main grid supply interruption is captured by defining a set of islanding scenarios with various start times and durations. A reliable operation of the microgrid in islanding scenarios would ensure resiliency. The market price forecast error is overlooked in the model since it would not appear in the microgrid resilient operation problem.

A microgrid must have sufficient online capacity during its normal operations to be able to supply the loads in resilient operations. If the mismatch is not zero, i.e. feasible islanding cannot be obtained, the normal operations solution is revised. The revision of the normal operation solution is performed via three actions, as follows: (i) changing the commitment of dispatchable units and the schedule of the energy storage, (ii) changing the schedule of adjustable loads, and (iii) load curtailment. The change in the commitment of dispatchable units and the schedule of the energy storage is considered the first action since it may increase the operation cost but does not cause any inconvenience for consumers. If this change does not result in a feasible islanding, the model would impose changes to the schedule of adjustable loads. This change would enable shifting away adjustable loads from islanding hours and, accordingly, reduce mismatches. The inconvenience experienced by consumers because of this change is penalized under the objective of the normal operations problem. If, after these revisions, a feasible islanding still cannot be obtained, the microgrid master controller would curtail loads as a last resort. The load curtailment is performed with the objective of removing power mismatches and matching available generation with the load. Loads are curtailed based on the load criticality criterion, i.e. less important loads are curtailed first, and if needed, more critical loads are considered for curtailment. These changes are governed by forming resiliency cuts in the resilient operation problem and sending them back to the normal operation problem for subsequent iterations. The final solution is obtained when all mismatches are zero and all islanding scenarios are guaranteed to be feasible.

3.3   An Economic Model for Microgrid Resilience

The normal operation problem formulation is as follows:

$$\text{Min} \sum_{t} \sum_{i \in G} [F_i(P_{it})I_{it} + SU_{it} + SD_{it}] + \sum_{t} \rho_t P_{M,t} + \sum_{d \in D} K_d \Delta_d \qquad (3.1)$$

s.t.

$$\sum_{i} P_{it} + P_{M,t} = \sum_{d} D_{dt} \qquad \forall t \qquad (3.2)$$

$$-P_M^{max} \leq P_{M,t} \leq P_M^{max} \qquad \forall t \qquad (3.3)$$

$$P_i^{min} I_{it} \leq P_{it} \leq P_i^{max} I_{it} \qquad \forall i \in G, \forall t \qquad (3.4)$$

$$P_{it} - P_{i(t-1)} \leq UR_i \qquad \forall i \in G, \forall t \qquad (3.5)$$

$$P_{i(t-1)} - P_{it} \leq DR_i \qquad \forall i \in G, \forall t \qquad (3.6)$$

$$T_i^{on} \geq UT_i (I_{it} - I_{i(t-1)}) \qquad \forall i \in G, \forall t \qquad (3.7)$$

$$T_i^{off} \geq DT_i (I_{i(t-1)} - I_{it}) \qquad \forall i \in G, \forall t \qquad (3.8)$$

$$P_{it} \leq P_{it}^{dch,\,max} u_{it} - P_{it}^{ch,\,min} v_{it} \qquad \forall i \in S, \forall t \qquad (3.9)$$

$$P_{it} \geq P_{it}^{dch,\,min} u_{it} - P_{it}^{ch,\,max} v_{it} \qquad \forall i \in S, \forall t \qquad (3.10)$$

$$u_{it} + v_{it} \leq 1 \qquad \forall i \in S, \forall t \qquad (3.11)$$

$$C_{it} = C_{i(t-1)} - P_{it} \qquad \forall i \in S, \forall t \qquad (3.12)$$

$$0 \leq C_{it} \leq C_i^{max} \qquad \forall i \in S, \forall t \qquad (3.13)$$

$$T_i^{ch} \geq MC_i (u_{it} - u_{i(t-1)}) \qquad \forall i \in S, \forall t \qquad (3.14)$$

$$T_i^{dch} \geq MD_i (v_{it} - v_{i(t-1)}) \qquad \forall i \in S, \forall t \qquad (3.15)$$

$$D_{dt}^{min} z_{dt} \leq D_{dt} \leq D_{dt}^{max} z_{dt} \qquad \forall d \in D, \forall t \qquad (3.16)$$

$$\sum_{t \in [\alpha_d, \beta_d]} D_{dt} = E_d \qquad \forall d \in D \tag{3.17}$$

$$T_d^{\mathrm{on}} \geq MU_d \left(z_{dt} - z_{d(t-1)} \right) \qquad \forall d \in D, \forall t \tag{3.18}$$

$$\Delta_d = \left(\beta_d^{\mathrm{new}} - \alpha_d^{\mathrm{new}} \right) - \left(\beta_d - \alpha_d \right) \qquad \forall d \in D \tag{3.19}$$

The resilient operation problem formulation is, as follows:

$$\mathop{\mathrm{Max}}_{U} \mathop{\mathrm{Min}}_{P} \ w_s = \sum_t \left(SL_{1,ts} + SL_{2,ts} \right) \tag{3.20}$$

subject to (3.4)–(3.6), (3.9)–(3.10), (3.12)–(3.13), (3.16)–(3.17) for each scenario s, and the following additional constraints:

$$\sum_i P_{its} + SL_{1,ts} - SL_{2,ts} = \sum_d D_{dts} \qquad \forall t \tag{3.21}$$

$$I_{its} = \hat{I}_{it} \quad \lambda_{its} \qquad \forall i \in G, \forall t \tag{3.22}$$

$$u_{its} = \hat{u}_{it} \quad \mu_{its}^{\mathrm{dch}} \qquad \forall i \in S, \forall t \tag{3.23}$$

$$v_{its} = \hat{v}_{it} \quad \mu_{its}^{\mathrm{ch}} \qquad \forall i \in S, \forall t \tag{3.24}$$

$$z_{dts} = \hat{z}_{dt} \quad \pi_{dts} \qquad \forall d \in D, \forall t \tag{3.25}$$

$$\left| P_{its} - \hat{P}_{it} \right| \leq RR \qquad \forall i \in G, \forall t \tag{3.26}$$

$$-P_M^{\mathrm{max}} U_{ts} \leq P_{M,ts} \leq P_M^{\mathrm{max}} U_{ts} \qquad \forall t \tag{3.27}$$

where b is index for energy storage, d is index for loads, i is index for DERs, s is index for scenarios, t is index for time, $\wedge$ is index for calculated variables, D is set of adjustable loads, G is set of dispatchable units, P is set of primal variables, S is set of energy storage systems, U is set of uncertain parameters, W is set of non-dispatchable units, DR is ramp-down rate, DT is minimum downtime, E is load total required energy, $F(.)$ is generation cost, K_d is inconvenience penalty factor, MC is minimum charging time, MD is minimum discharging time, MU is minimum operating time, RR is permissible power adjustment, U is outage state of main grid line/islanding state, UR is ramp-up rate, UT is minimum uptime, α, β are specified start and end times of adjustable load, ρ is market price, C is energy storage system state of charge (SOC), D is load demand, I is commitment state of dispatchable unit, P is DER output power, P_M is main grid power, SD is shutdown cost, SL_1, SL_2 are slack variables, SU is startup cost, T^{ch} is number of successive charging hours, T^{dch} is number of successive discharging hours, T^{on} is number of successive ON hours, T^{off} is number of successive OFF hours, u is energy storage system discharging state, v is energy storage system charging state, w is power mismatch, z is adjustable load state, λ, μ, π are dual variables, and Δ_d is deviation in adjustable load operating time interval.

The objective of the normal operation problem is to minimize the microgrid operations cost, including the operations cost of dispatchable units, the cost of power transfer from the main grid, and the inconvenience cost realized by consumers (3.1). The cost of power transfer from the main grid could be positive or negative, depending on the direction of flow in the transmission line connecting the microgrid to the main grid. A negative cost, which represents a power export to the main grid, appears as an economic benefit for the microgrid. The inconvenience cost represents the penalty in scheduling adjustable loads outside the time intervals specified by consumers. The constant penalty factor, K_d, is used to prioritize the loads with regard to sensitivity when operating within the specified time intervals, where a higher value for K_d represents a less flexible load in terms of operating time interval adjustments. The value for K_d is selected reasonably higher than the generation cost of units and the market price.

The load balance constraint (3.2) ensures that the sum of power generated by DERs and power from the main grid would match the hourly load. The power transfer with the main grid is limited by flow limits of the line connecting the microgrid to the main grid (3.3). The dispatchable unit generation is subject to minimum and maximum generation capacity limits (3.4), ramp-up and ramp-down rate limits (3.5)–(3.6), and minimum up and downtime limits (3.7) and (3.8). The unit commitment state, I_{it}, is one when unit is committed and is zero otherwise. The energy storage power could be positive (discharging), or negative (charging). In either case, the energy storage power is limited by minimum and maximum power constraints (3.9)–(3.11). The energy storage state of charge (SOC) is calculated based on the amount of charged/discharged power (3.12) and restricted with capacity limits (3.13). Minimum charging and discharging time limits, i.e. the minimum number of consecutive hours that the energy storage must maintain its operational mode, are also considered (3.14) and (3.15). Adjustable loads are subject to minimum and maximum rated powers (3.16) and would consume the required energy to complete an operating cycle in the time interval specified by the consumer (3.17). Certain loads may be subject to minimum operating time (3.18), which is the number of consecutive hours that a load must consume power once it is switched on. Constraint (3.19) reflects the change in adjustable loads schedules in the normal operation problem, where $\left[\alpha_d^{\text{new}}, \beta_d^{\text{new}}\right]$ represents the new operating time interval, which is ensured to be larger than the initially specified time interval, i.e. $\beta_d^{\text{new}} \geq \beta_d$ and $\alpha_d^{\text{new}} \leq \alpha_d$.

Once the normal operation problem solution is obtained, the resilient operation problem will be solved. The objective of the resilient operation problem for an islanding scenario s is to minimize the power mismatches, as in (3.20). Power balance Eq. (3.21) encompasses slack variables SL_1 and SL_2, which act as virtual generation and virtual load, respectively. A nonzero value for either of these variables denotes a power mismatch in the microgrid's resilient operation. The commitment of dispatchable units and the schedule of energy storage and adjustable loads are obtained from the normal operation problem. The given variables are replaced with local variables for obtaining associated dual multipliers (3.22)–(3.25) and further enable forming the resiliency cut. The permissible change in dispatchable unit output from the normal operation to resilient operation is represented by (3.26).

To capture the load and non-dispatchable generation forecast uncertainties, robust programming is employed, in which the worst-case solution of the resilient operation problem is to be found over uncertainty set U. To find this robust solution, the objective (3.20) is maximized over the uncertainty set to find the worst-case solution of the mismatch minimization problem. The obtained max–min problem is complex to solve, and an efficient way to solve it is by finding the dual problem of the inner minimization problem and combining it with the outer maximization problem. The worst-case solution will be obtained at extreme points of uncertain parameters [23, 24]. In the proposed resilient operation problem, however, the extreme points of uncertain parameters, i.e. non-dispatchable generation and load, could be simply obtained. A higher load and a lower non-dispatchable generation will result in a higher mismatch; thus, the worst-case solution would be obtained when the non-dispatchable generation is at its lower uncertainty bound and the load is at its upper uncertainty bound. The power balance constraint is accordingly replaced with (3.28), where inserted bars represent the upper bound and the lower bound of the load and non-dispatchable generation forecast uncertainty intervals, respectively.

$$\sum_{i \in G} P_{it} + \sum_{i \in W} \underline{P}_{it} + SL_{1,t} - SL_{2,t} = \sum_{d \in D} D_{dt} + \sum_{d \notin D} \overline{D}_{dt} \quad \forall t \tag{3.28}$$

To capture the uncertainty due to main grid supply interruptions, a binary outage state is included in the main grid power constraint (3.27), and accordingly, islanding scenarios are defined. A zero value for the outage state would model the microgrid islanding as it imposes a value of zero on the main grid power. Each islanding scenario would start at a different hour and would last for the maximum predicted interruption time. For example, if the incident is predicted to impact the main grid between hours $t + 1$ and $t + m$ and the estimated maximum repair and restoration time is T hours, a total of m scenarios will be considered for islanding, and each would last for T hours. The binary outage state in (3.27) is determined offline based on islanding scenarios. The feasible islanding is examined in all scenarios, and if there is any mismatch, the normal operation problem solution would be revised using the resiliency cut (3.29). This cut results in a change in the unit commitment states, energy storage schedule, and adjustable loads schedules based on resiliency considerations. The iterative process continues until power mismatches in all islanding scenarios reach zero.

$$
\begin{aligned}
\hat{w}_s + \sum_{i \in G} \lambda_{its}\left(I_{it} - I_{its}\right) + \sum_{i \in S} \mu_{its}^{dch}\left(u_{it} - u_{its}\right) \\
+ \sum_{i \in S} \mu_{its}^{ch}\left(v_{it} - v_{its}\right) + \sum_{i \in D} \pi_{dts}\left(z_{dt} - z_{dts}\right) \le 0
\end{aligned}
\tag{3.29}
$$

It is probable that after a certain number of iterations and revising the unit commitment states, energy storage schedule, and adjustable loads schedules, a feasible islanding cannot be achieved, and the power mismatch still remains. In this case, the microgrid master control will, therefore, curtail loads. This action is considered as the last resort since it causes a significant inconvenience for microgrid consumers. The microgrid master controller will simply curtail the load, equal to the power mismatch between the available generation and load, to achieve the feasible islanding and ensure resiliency. The curtailment, however, would be performed based on the load criticality, in which more critical loads have a lower priority for curtailment. Once curtailed, the problem converges, and there would be no need to perform further iterations.

3.4 Case Study

The proposed resiliency-oriented microgrid optimal scheduling model is applied to a microgrid test system. The data for generating units, energy storage systems, and adjustable loads, as well as the forecasted values of microgrid hourly fixed load, non-dispatchable units' generation, and market price, are borrowed from [22]. A constant penalty factor of $100/h, for every hour deviation from adjustable loads specified start and end times, is considered. A 24-hour scheduling horizon is considered for studies, assuming that the incident occurs, and the possible damages are repaired, within this horizon. Any other scheduling horizon can be selected based on the microgrid master controller's prediction of the main grid restoration time. Dispatchable units' commitments and energy storage charging/discharging schedules will be determined in the normal operation problem and remain unchanged in the resilient operation problem. It is assumed that microgrid components are not subject to outages during the scheduling horizon. The resiliency-oriented microgrid optimal scheduling is studied considering an uncertain main grid supply interruption, as well as uncertain load and generation forecasts. It is predicted that the incident will result in main grid supply interruption at noon.

Based on forecasts, the damage to the main grid will be repaired and the supply will be restored in less than seven hours from the time of incident. A forecast error of $\pm 20\%$ for non-dispatchable generation and $\pm 10\%$ for load is considered.

Case 1—Impact of uncertainties on microgrid scheduling: The resiliency-oriented microgrid optimal scheduling problem is solved for two cases. In the first case, the microgrid master controller assumes that the incident would occur exactly at noon, i.e. the time of incident is known, and then schedules microgrid resources for a seven-hour resilient operation. In the second case, the microgrid master controller considers a two-hour uncertainty in the time of incident, so it would solve the resilient operation problem for five different islanding scenarios, from 10 a.m. to 2 p.m., each lasting for seven hours. The problem is implemented on a 2.4-GHz personal computer using CPLEX 11.0 [25].

Case 1(a): The resiliency-oriented microgrid optimal scheduling results in a total operation cost of $11,855, considering known incident time and duration. In the resiliency mode, the power transfer from the main grid is zero; therefore, sufficient capacity is committed in the normal operation to enable a quick switch to resilient operation without interruption in load supply. Accordingly, microgrid components are scheduled to be able to supply the load for seven consecutive hours. In the first iteration, the solution of the normal operation problem commits only two dispatchable units 1 and 2, which results in 8.61 MWh power mismatch in resilient operation. However, additional commitment of units 3 and 4, revising the adjustable loads schedules, and revising the energy storage discharging schedule at subsequent iterations, reduces the total mismatch to 3.57 MWh. This mismatch could not be further reduced; thus load is curtailed.

In the obtained solution, dispatchable units 3 and 4 are not economical and are committed due to resiliency considerations; therefore, they would be dispatched at their minimum generation capacities. The energy storage is discharged at a slow rate for seven hours to cooperate with the microgrid's resilient operation when the available unit capacity cannot completely supply the local load. The schedule of adjustable loads is changed, where most of adjustable loads are moved toward end of the day and are partially scheduled in the specified time horizon. These changes result in reduced inconvenience for consumers and correspond to lower rate hours in normal operation. The obtained solution indicates that by committing additional units, changing the energy storage schedule, and revising adjustable loads schedules, the load curtailments in case of main grid supply interruption and under uncertain load and generation could be significantly reduced.

The sensitivity of solution with respect to load and non-dispatchable unit generation forecast uncertainties is analyzed and depicted in Figures 3.3 and 3.4, respectively. In Figure 3.3, the generation forecast error is fixed at 20%, while the load forecast error is changed from zero to 10%. In Figure 3.4, the load forecast error is fixed at 10%, while the generation forecast error is changed from 0

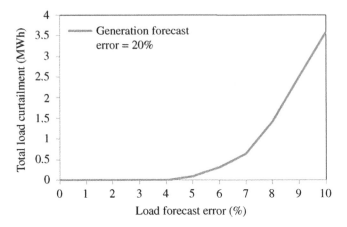

Figure 3.3 Total load curtailment as a function of load forecast error.

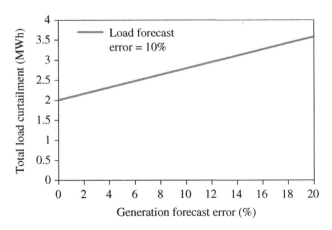

Figure 3.4 Total load curtailment as a function of generation forecast error.

to 20%. The results advocate that the microgrid scheduling solution depends significantly on load forecast errors. By increasing the generation forecast error from 0 to 10%, the load curtailment is increased by 0.7 MWh, while the related increase in the load forecast error is more than 3.5 MWh. Furthermore, the load curtailment increases almost linearly by increasing the generation forecast error, while this relationship for increase in the load forecast error is exponential. The result suggests that a more accurate load forecast has a more significant role in the microgrid's resilient operations compared to generation forecasts.

Case 1(b): In the second case, the problem is solved for an uncertain start time of the incident. Since the incident is forecast to occur at noon with a two-hour uncertainty, five scenarios are considered in the resilient operation problem, each lasting for seven hours. Scenarios 1–5 will consider islanding from 10 a.m., 11 a.m., 12 p.m., 1 p.m., and 2 p.m., respectively. The initial normal operation schedule results in an average of 11.2 MWh in all scenarios in the first iteration. By forming the resiliency cut, the mismatch will be reduced in subsequent iterations. After six iterations a feasible islanding in scenarios is not guaranteed, while additional units 3 and 4 are committed, and adjustable loads are scheduled outside their specified operating time interval. As an option of last resort, the microgrid master controller partially curtails the load in each scenario; thus, the load and generation in all scenarios would match. The total generation cost in this case is $12,087, which includes generation cost of local DERs, energy purchases from the main grid, and the consumer inconvenience cost. The final solution includes an average of 3.7 MWh of load curtailment in scenarios, with the lowest in scenario 1 (i.e. 0.89 MWh) and the highest in scenario 5 (i.e. 6.50 MWh). The higher load curtailment in scenario 5 compared to scenario 1 is due to the reduced available charge of the energy storage at hours 20–21. When reaching these times, the energy storage is completely depleted. Moreover, the generation of renewable resources is zero. Thus, the microgrid master controller is required to curtail more loads at these hours to maintain the supply and load balance.

A comparison between these two cases demonstrates that the solution of the second case results in a higher operation cost and inconvenience for consumers as well as a higher expected load curtailment; however, this solution is more resilient than that in the first case. The solution of the second case ensures a robust microgrid operation against uncertain main grid supply interruptions that occurred in any of hours between 10 a.m. and 2 p.m. To further elaborate the resiliency of the second case solution, assume that the main grid supply is interrupted at hour 10 and will last for 7 hours. In this situation, the solution of the second case would not change, as it already considers this interruption and accordingly schedules microgrid resources for resilient operation. However, the solution of the first case would result in significant load curtailments. If scenario 1 occurs, i.e. the interruption starts at hour 10, the microgrid would have to curtail more than 17 MWh to balance load and supply during resilient operation. Bulk of the curtailment occurs at hours 10 and 11, when the microgrid has not scheduled generation units as well as the energy storage to supply the load in case of islanding. This comparison advocates that considering an uncertain main grid supply interruption time may result in a higher operation cost but would be more robust compared to the case when this uncertainty is not taken into account.

Solving the optimal microgrid scheduling, without resiliency considerations, the total operating cost is obtained as $11,183. It shows that the resiliency-oriented scheduling has resulted in 6 and 8% increase in the total operation cost in the first and second cases, respectively. This cost could be considered as the cost of resiliency, which is added to the microgrid operation cost for guaranteeing a reliable supply of loads during main grid supply interruptions. When compared with the amount of avoided microgrid curtailment, this cost is insignificant, which shows the viability of resiliency-oriented microgrid scheduling.

Case 2—Impact of permissible power adjustment on microgrid scheduling: The impact of permissible power adjustments on microgrid scheduling solutions is studied in this case. Uncertainty in load, renewable generation, and main grid supply interruption time and duration is considered as that in Case 1(b). The permissible power adjustment would restrict the change in dispatchable units' output when the microgrid switches from normal operation to resilient operation. A higher permissible adjustment provides more flexibility in scheduling for resilient operation, results in less load curtailment, and reduces the operation cost. Results of this analysis, based on operation costs, are depicted in Figure 3.5, in which permissible power adjustment is presented as a percentage of associated dispatchable unit ramp rate. A zero permissible power adjustment represents that the dispatchable units' output cannot be changed when switching to resilient operation. This task is analogous to considering unit power output as a preventive action in dealing with resiliency rather than a corrective action. Although this action is much simpler to schedule and perform than a corrective action, the operation cost will be increased, as shown in Figure 3.5.

Microgrids are viable technologies for improving power system resiliency by promoting the dispersion of power resources and islanding. An efficient

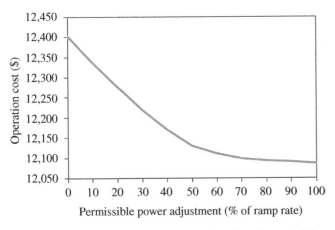

Figure 3.5 Microgrid operation cost as a function of permissible power adjustment.

mathematical modeling of the microgrid optimal scheduling problem based on resiliency considerations, however, is required to deliver expected benefits. Specific features of the proposed resiliency-oriented microgrid optimal scheduling model are listed as follows:

- *Least-cost normal operations:* The microgrid optimal scheduling determines the operation of generating units, energy storage systems, and adjustable loads, along with the main grid power transfer, to minimize the cost of supplying local loads in normal power system operation.
- *Resiliency consideration:* Sufficient DER capacity is scheduled to enable a seamless islanding. If required, adjustable loads schedules are revised, and additional loads are curtailed to enable resilient operations.
- *Uncertainty consideration:* Forecast errors involved in load and non-dispatchable generation forecasts, as well as uncertain main grid supply interruption time and duration, are captured in the microgrid resilient operation using a robust optimization method and via worst-case analysis.
- *Consumer convenience:* The consumer's decisions in scheduling adjustable loads are not changed unless it is required to obtain a feasible islanding solution. The changes, however, are penalized to reduce the inconvenience for consumers and reflect the load schedule outside specified operating time intervals. Furthermore, additional load curtailments are performed based on load criticality as a last resort for removing power mismatches.
- *Operational flexibility:* The proposed model provides an efficient method for the microgrid master controller to employ the available resources to address resiliency needs.

3.5 Summary

One of complementary value propositions of microgrids is to improve power system resiliency via local supply of loads and curtailment reduction. This subject was investigated in this chapter by proposing a resiliency-oriented microgrid optimal scheduling model. The proposed model considers prevailing uncertainties in load, generation, and main grid supply interruption time and duration. This model aims at minimizing the microgrid load curtailment by efficiently scheduling available resources when supply of power from the main grid is interrupted for an extended period of time. The problem is decomposed to normal operation and resilient operation problems. The normal operation problem solution, i.e. unit commitment states, energy storage schedules, and adjustable loads schedules, is employed in the resilient operation problem to examine microgrid capability in supplying local loads during main grid supply interruption. The schedule is revised via resiliency cuts if a zero mismatch is not obtained. Prevailing operational uncertainties in load, non-dispatchable generation, and the main grid supply interruption time and duration are considered and captured using a robust optimization method. The feasibility of resilient operation was ensured via three actions, which respectively revised the unit commitments and energy storage schedules, revised adjustable loads schedules, and curtailed loads. Main grid supply interruption uncertainty was captured via islanding scenarios. Mixed integer programming was used to model the normal operation problem, and linear programming was used to model the resilient operation problem. The final solution, which is obtained in an iterative manner, is economically optimal, guarantees robustness against prevailing operational uncertainties, and supports a quick islanding with minimum consumer inconvenience and load curtailment. A set of case studies demonstrate the effectiveness of the proposed resiliency-oriented microgrid optimal scheduling model.

References

1 A. Flueck and Z. Li, "Destination perfection," *IEEE Power Energ. Mag.*, vol. 6, no. 6, pp. 36–47, Nov./Dec. 2008.

2 M. Shahidehpour and J. Clair, "A functional microgrid for enhancing reliability, sustainability, and energy efficiency," *Electr. J.*, vol. 25, pp. 21–28, Oct. 2012.

3 S. Bahramirad, W. Reder, and A. Khodaei, "Reliability-constrained optimal sizing of energy storage system in a microgrid," *IEEE Trans. Smart Grid*, vol. 3, no. 4, pp. 2056–2062, Dec. 2012.

4 I. Bae and J. Kim, "Reliability evaluation of customers in a microgrid," *IEEE Trans. Power Syst.*, vol. 23, no. 3, pp. 1416–1422, Aug. 2008.

5 A. G. Tsikalakis and N. D. Hatziargyriou, "Centralized control for optimizing microgrids," *IEEE Trans. Energy Conv.*, vol. 23, no. 1, pp. 241–248, 2008.

6 Federal Energy Management Program, "Using Distributed Energy Resources, a How-to Guide for Federal Facility Managers," DOE/GO-102002-1520, US Department of Energy, May 2002.

7 A. Khodaei and M. Shahidehpour, "Microgrid-based co-optimization of generation and transmission planning in power systems," *IEEE Trans. Power Syst.*, vol. 28, no. 2, pp. 1582–1590, May 2013.

8 S. Chowdhury, S. P. Chowdhury, and P. Crossley, "Microgrids and Active Distribution Networks," *IET Renewable Energy Series*, 2009.

9 Federal Energy Management Program, "Using Distributed Energy Resources, a How-to Guide for Federal Facility Managers," DOE/GO-102002-1520, US Department of Energy, May 2002.

10 M. Shahidehpour, "Role of smart microgrid in a perfect power system," in *IEEE Power and Energy Society General Meeting*, Minneapolis, MN, 2010.

11 V. Meyer, C. Myres, and N. Bakshi. "The vulnerabilities of the power-grid system: renewable microgrids as an alternative source of energy," *J. Bus. Contin. Emer. Plan.*, vol. 4, no. 2, pp. 142–153, 2010.

12 P. Agrawal "Overview of DOE Microgrid Activities," in *Symposium on Microgrid*, Montreal, QC, Canada, vol. 23, June 2006.

13 A. G. Tsikalakis and N. D. Hatziargyriou, "Operation of microgrids with demand side bidding and continuity of supply for critical loads," *Eur. Trans. Electr. Power*, vol. 21, no. 2, pp. 1238–1254, Mar. 2011.

14 S. Rahman, "Framework for a Resilient and Environment-Friendly Microgrid with Demand-Side Participation," in *IEEE Power and Eng. Soc. Gen. Meeting - Conversion and Delivery of Electrical Energy in the 21st Century*, Pittsburgh, PA, 2008.

15 P. Li, P. Degobert, B. Robyns, and B. Francois. (2008). "Implementation of Interactivity Across a Resilient Microgrid for Power Supply and Exchange with an Active Distribution Network," IET Digital Library.

16 F. O. Resende, N. J. Gil, and J. A. P. Lopes, "Service restoration on distribution systems using multi-MicroGrids," *Eur. Trans. Electr. Power*, vol. 21, no. 2, pp. 1327–1342, Mar. 2011.

17 C. M. Colson, M. H. Nehrir, and R. W. Gunderson, "Distributed multi-agent microgrids: a decentralized approach to resilient power system self-healing," in *4th International Symposium on Resilient Control Systems*, Boise, ID, pp. 83–88, 2011.

18 J. Hurtt and L. Mili, "Residential microgrid model for disaster recovery operations," in *IEEE Grenoble Conference*, Grenoble, France, 2013.

19 C. Gouveia, J. Moreira, C. L. Moreira, and J. A. Pecas Lopes, "Coordinating storage and demand response for microgrid emergency operation," *IEEE Trans. Smart Grid*, vol. 4, pp. 1898–1908, 2013.

20 X. Xu, J. Mitra, N. Cai, and L. Mou, "Planning of reliable microgrids in the presence of random and catastrophic events," *Int. Trans. Electr. Energy Syst.*, vol. 24, pp. 1151–1167, 2013.

21 S. Cano-Andrade, M. R. von Spakovsky, A. Fuentes, C. Lo Prete, B. F. Hobbs, and L. Mili, "Multi-objective optimization for the sustainable-resilient synthesis/design/ operation of a power network coupled to distributed power producers via microgrids," in *Proceedings of the ASME 2012 International Mechanical Engineering Congress and Exposition. Volume 6: Energy, Parts A and B*, Houston, TX, pp. 1393–1408, Nov. 9–15, 2012.

22 A. Khodaei, "Microgrid optimal scheduling with multi-period islanding constraints," *IEEE Trans. Power Syst.*, vol. 29, pp. 1383–1392, 2014.

23 A. Ben-Tal, L. El Ghaoui, and A. Nemirovski, "Robust Optimization," Princeton University Press, 2009.

24 A. Thiele, T. Terry, and M. Epelman, "Robust Linear Optimization with Recourse," Rapport Technique, pp. 4–37, 2009.

25 ILOG CPLEX, "ILOG CPLEX Homepage 2009," [Online]. Available: www.ilog. com, 2009.

4

Community Microgrid Operations Management

4.1 Principles of Community Microgrids

Community microgrids represent a novel application of this technology with the primary objective of providing reliable power supply to residential neighborhoods. Community microgrids will provide a viable solution for consumers who cannot solely rely on the supply of power from the main grid due to a variety of reasons, including the criticality of their load, sensitivity to power disruption, and economic objectives that can be achieved by utilizing the local power generation capacity, among others. Moreover, community microgrids represent practical alternatives for vertically integrated electric utility companies in upgrading power delivery facilities and address the ever-increasing need for higher levels of quality, reliability, and efficiency in power supply. The transition from the conventional utility grid to smart community microgrids and the enhanced utilization of distributed energy resources (DERs) and controllable loads will offer significant opportunities, including increased energy efficiency, enhanced conservation levels, and lowered greenhouse gas emissions, while lowering the stress level on congested transmission lines. These microgrids, in addition, can address the growing use of distributed generation by residential customers, address the growing need for enhanced power quality and reliability that can be realized from the increased use of digital devices and voltage-sensitive loads, and the emerging class of loads in the distribution networks, including plug-in vehicles and energy storage.

Community microgrids can be considered an attractive alternative to a vertically integrated, centralized generation and bulk transmission of electricity given the significant share of power usage by residential customers [1, 2]. However, empirical studies on the real-world impacts of community microgrid deployment are limited, even though microgrids have been extensively investigated in the literature [3–14]. The majority of microgrid studies and development efforts are devoted

The Economics of Microgrids, First Edition. Amin Khodaei and Ali Arabnya.
© 2024 The Institute of Electrical and Electronics Engineers, Inc.
Published 2024 by John Wiley & Sons, Inc.

to campus microgrids (conceivably due to higher adoption rate of new technologies on campuses, advocacy by their faculty with an active research area in the field, as well as the availability of in-house expertise in academic environments), remote microgrids (primarily due to providing a quick and efficient alternative to generation and transmission infrastructure upgrades to deploy microgrids), and military microgrids (primarily due to extensive government support and proved viability in providing self-sustaining small power grids required to address the mission-critical energy needs of military bases). More recently, commercial and industrial consumers are considering microgrid deployments as a means to increase power supply reliability and prevent significant business interruptions as a result of power disruptions or power quality issues. Major obstacles to the rapid and widespread deployment of community microgrids include, but are not limited to the relatively high amount of initial capital expenditure for the deployment of microgrids, lack of adequate consumer knowledge on potential economic and reliability impacts of distributed generation and load scheduling strategies, as well as ownership structure, governance systems, and regulatory issues. However, as electric utilities are embracing microgrids as part of their business imperative, it is expected to observe a higher rate of utility-deployed community microgrids, going forward.

A major difference between community microgrids and other types of microgrids – that significantly adds to the complexity of the community microgrid operations and control – is in the design requirements of their master controller. Microgrid master controller is a decision-making software installed on a high-performance computer that acts as the brain of the microgrid. Microgrid master controller determines the hourly scheduling of DERs in the microgrid (e.g. natural gas, fuel cell, rooftop solar panels, wind turbines, and energy storage, among others), interactions with the utility grid, decision to switch between grid-connected and islanded modes, frequency regulation and voltage control, and any decisions on load curtailment/shifting, all based on economic and reliability considerations. However, unlike other types of microgrids, the master controller for the community microgrid does not have full access and authority to monitor and control individual loads and DERs in different building blocks of the microgrid. For example, consider a police station, a hospital, or a library with a variety of critical and sensitive loads, which need to be locally operated, and thus, would not permit external control by the microgrid master controller. A local building controller to locally manage these resources based on price signals and accordingly maintain the required privacy levels is more desirable in this case. Therefore, an important challenge that a master controller faces is to control and optimally schedule microgrid load, utilize available resources, and manage transactions with the main grid, while having limited authority and control over microgrid DERs and loads.

This chapter discusses the community microgrid optimal operations scheduling as a tertiary control mechanism performed by the microgrid master controller. The traditional resource scheduling models cannot be applied to microgrids considering microgrid's special characteristics including the considerable size of non-dispatchable renewable energy resources and energy storage compared to local loads, short distance between loads and DERs, the removal of network congestion, the considerable size of adjustable loads with the ability to respond to electricity prices and control signals, the connection to the main grid, which acts as a backup generation/load, and islanding capability. On top of these, the limited control of the community microgrid master controller on local loads and DERs must be considered. We adopt a hybrid architecture for the master controller, which takes advantage of both centralized and distributed schemes. A centralized scheme ensures secure microgrid operations and is more suitable for the application of optimization techniques by collecting loads and DERs information and centrally optimizing the microgrid operation [15–21]. Distributed scheme relies on individually controlled and scheduled intelligent components (i.e. agents) with the ability of discrete decision-making, where the optimal schedule is obtained using iterative data transfers among agents [22–24]. Islanding considerations are an inherent part of these schemes, which are considered in literature by exploiting reserve constraints [25], energy storage [26–28], and a combination of DERs and adjustable loads [21]. The objective of the proposed hybrid architecture is to minimize the day-ahead operations cost in grid-connected mode and maximize the reliability in islanded mode, while maintaining the privacy of individual DERs and loads. An iterative model links the decisions of the master controller with individual decisions of buildings. While the proposed model is specifically developed for community microgrids, however, without loss of generality, it is also applicable to other types of microgrids.

4.2 Economic Variables in Community Microgrid Operations

Community microgrid loads and DERs can be categorized into two types: *communal* and *individual,* which can come from operations and control mechanisms. Communal DERs are installed by the microgrid developer and controlled by the master controller to benefit the entire microgrid. These DERs are utilized to facilitate reliable operations in the islanded mode as well as enhance economical operations in the grid-connected mode. An example is community energy storage. Communal loads are those that will be used by the entire community, and the associated costs will be distributed among customers, such as streetlights. Individual DERs, on the other hand, belong to a specific customer (building) within the microgrid, hence operated and controlled by the associated local controller. Examples include backup distributed generation units in hospitals or police stations or rooftop solar panels of residential customers, among others. Individual loads are loads (both fixed and adjustable) for a customer within the community microgrid that can be locally scheduled by the local controller rather than centralized and direct control by the microgrid master controller.

The main variables in the optimal scheduling of microgrid operations (for both communal and individual), include loads, generating units, and energy storage. Loads are categorized into two types: fixed (which cannot be altered under normal operating conditions) and adjustable (which are responsive to price variations and/or control signals). Generation units are categorized into two groups: dispatchable (that can be controlled by the master or local controller and are subject to technical constraints) and non-dispatchable (which cannot be controlled since the input variable is uncontrollable). Energy storage is utilized to ensure microgrid generation adequacy, load shifting, and seamless islanding capabilities. Figure 4.1 depicts the structure of the proposed community microgrid master controller, representing the interaction of the master controller with communal and individual DERs and loads. The proposed hybrid model leverages the merits of both centralized and distributed control schemes. The only available control mechanism is the price signal, which will be sent to local controllers from the master controller. Each local controller will schedule its DERs and loads and send back the net load data (i.e. building total load minus local generation) to the master controller. The net load can be positive, representing a consumer, or negative, representing a generator. The master controller schedules communal DERs and loads, using direct control signals, based on economic and reliability considerations. Figure 4.2 depicts the workflow of the proposed scheduling scheme. As shown, the initial price data, which is obtained

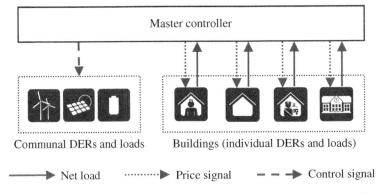

Communal DERs and loads   Buildings (individual DERs and loads)

——▶ Net load   ·········▶ Price signal   — — ▶ Control signal

Figure 4.1 Structure of the proposed community microgrid master controller.

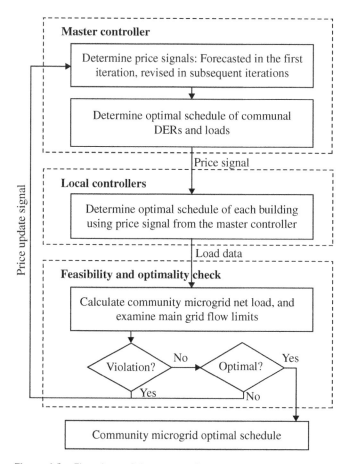

Figure 4.2 Flowchart of the proposed community microgrid master controller design.

from the utility or forecasted by the master controller, will be revised in each iteration to efficiently alter the load schedule and meet the corresponding operational constraints.

The islanding requirements will be further included in the optimal scheduling model. The microgrid must be able to switch to the islanded mode at any given time in response to uncertain disturbances in the main grid (i.e. time and duration of the disturbance are not known in advance, but the microgrid is ready to respond to the event). Microgrid resources can be scheduled accordingly to supply loads in grid-connected mode and enable a seamless switching to the islanded mode while supplying local loads. Once the disturbances are eliminated, the microgrid can be resynchronized with the main grid. Without loss of generality, a 24-hour planning horizon can be considered, which would enable master and local controllers to benefit from day-ahead market price forecasts provided by the utility company and to keep track of the energy storage system's daily charging and discharging cycles. Any other planning horizons can be used based on the master controller's discretion with minor changes in the proposed formulation.

4.3 An Economic Model for Community Microgrid Operations Management

4.3.1 Master Controller

The objective of the community microgrid master controller is to minimize the microgrid's total operations cost in grid-connected and islanded modes:

$$\text{Min} \sum_{s} \sum_{t} \sum_{i \in G} c_{is} P_{its} + \sum_{s} \sum_{t} \rho_t P_{M,ts} + \sum_{s} \sum_{t} v_t LS_{ts} \tag{4.1}$$

where s is an index for scenarios, t is an index for time, i is an index for DERs, G is a set of dispatchable generating units, c_{is} is generation marginal cost, P_{its} is DER output power, ρ_t is market price, $P_{M,ts}$ is main grid power, v_t is the value of lost load (VOLL), and LS_{ts} is load curtailment variable. The objective function is defined for operational scenarios represented by s, in which $s = 0$ denotes the grid-connected (base case) operation and $s > 0$ denotes the islanded (contingency) operation. The three terms in the objective represent the generation cost of dispatchable units, the cost of power transfer with the main grid, and the cost of load curtailment, respectively. A single-step price curve is considered for dispatchable units, which, without loss of generality, can be easily extended to a multi-step price curve. The generation price is defined for each scenario, where $c_{is} = c_{i0}, \forall s > 0$ if solution optimality in islanding scenarios is important, and $c_{is} = 0, \forall s > 0$ if only solution feasibility in islanding scenarios is important. The cost of main grid power transfer is calculated as the amount of transfer with the main grid multiplied by the market price at the point of common coupling. The market price is forecasted and published by the local electric utility company on a day-ahead basis, however, it can also be forecasted by the master controller. The cost of power transfer with the main grid can be positive, representing a cost, or negative, representing a profit for the microgrid, based on the direction of the flow (i.e. power import or export). The cost of load curtailment is defined as the VOLL times the amount of curtailed load. This objective models microgrid operations in both grid-connected and islanded modes. When grid-connected, load curtailment will be zero; hence, the objective represents a tradeoff between the generation cost and the cost of purchasing power from the main grid. When islanded, the cost of main grid power transfer will be zero – as the power cannot be exchanged with the main grid. The objective in this case represents a tradeoff between generation cost and the cost of load curtailment. The objective function is subject to the following constraints:

$$\sum_i P_{its} + P_{M,ts} + LS_{ts} = \sum_{d \in D_C} D_{dts} + \sum_{d \in D_B} D_{dts} \qquad \forall t, \forall s \tag{4.2}$$

$$-P_M^{\max} U_{ts} \leq P_{M,ts} \leq P_M^{\max} U_{ts} \qquad \forall t, \forall s \tag{4.3}$$

$$P_i^{\min} I_{it} \leq P_{its} \leq P_i^{\max} I_{it} \qquad \forall i \in G_C, \forall t, \forall s \tag{4.4}$$

$$P_{its} - P_{i(t-1)s} \leq UR_i \qquad \forall i \in G_C, \forall t, \forall s \tag{4.5}$$

$$P_{i(t-1)s} - P_{its} \leq DR_i \qquad \forall i \in G_C, \forall t, \forall s \tag{4.6}$$

$$T_i^{\text{on}} \geq UT_i\big(I_{it} - I_{i(t-1)}\big) \qquad \forall i \in G_C, \forall t, s = 0 \tag{4.7}$$

$$T_i^{\text{off}} \geq DT_i\big(I_{i(t-1)} - I_{it}\big) \qquad \forall i \in G_C, \forall t, s = 0 \tag{4.8}$$

$$|P_{its} - P_{it0}| \leq \Delta_i \qquad \forall i \in G_C, \forall t, \forall s \tag{4.9}$$

$$P_{its} \leq P_{it}^{\text{dch, max}} u_{it} - P_{it}^{\text{ch, min}} v_{it} \qquad \forall i \in S_C, \forall t, \forall s \tag{4.10}$$

$$P_{its} \geq P_{it}^{\text{dch, min}} u_{it} - P_{it}^{\text{ch, max}} v_{it} \qquad \forall i \in S_C, \forall t, \forall s \tag{4.11}$$

$$u_{it} + v_{it} \leq 1 \qquad \forall i \in S_C, \forall t, s = 0 \tag{4.12}$$

$$C_{its} = C_{i(t-1)s} - P_{its} u_{it}/\eta_i - P_{its} v_{it} \qquad \forall i \in S_C, \forall t, \forall s \tag{4.13}$$

$$C_i^{\min} \leq C_{its} \leq C_i^{\max} \qquad \forall i \in S_C, \forall t, \forall s \tag{4.14}$$

$$T_i^{\text{ch}} \geq MC_i\big(u_{it} - u_{i(t-1)}\big) \qquad \forall i \in S_C, \forall t, s = 0 \tag{4.15}$$

$$T_i^{\text{dch}} \geq MD_i\big(v_{it} - v_{i(t-1)}\big) \qquad \forall i \in S_C, \forall t, s = 0 \tag{4.16}$$

$$D_{dt}^{\min} z_{dt} \leq D_{dts} \leq D_{dt}^{\max} z_{dt} \qquad \forall d \in D_C, \forall t, \forall s \tag{4.17}$$

$$\sum_{t \in [\alpha_d, \beta_d]} D_{dts} = E_d \qquad \forall d \in D_C, \forall s \tag{4.18}$$

$$T_d^{\text{on}} \geq MU_d\big(z_{dt} - z_{d(t-1)}\big) \qquad \forall d \in D_C, \forall t, s = 0 \tag{4.19}$$

where ch is superscript for energy storage charging mode, dch is superscript for energy storage discharging mode, d is index for loads, D_B is set of building (individual) loads, D_C is set of communal loads, G_B is set of building (individual) dispatchable units, G_C is set of communal dispatchable units, S_C is set of communal energy storage systems, DR is ramp-down rate, DT is minimum downtime, E is load total required energy, MC is minimum charging time, MD is minimum discharging time, MU is minimum operating time, U is binary islanding indicator (1 in grid-connected mode, and 0 in islanded mode), UR is ramp-up rate, UT is minimum uptime, α is the specified start time of adjustable loads, β is the specified end time of adjustable loads, η is energy storage efficiency, Δ_d is generation adjustment limit, C is energy storage available energy, D is load demand, I is commitment state of the dispatchable unit (1 if committed, 0 otherwise), T^{ch} is number of successive charging hours, T^{dch} is number of successive discharging hours, T^{on} is number of successive ON hours, T^{off} is number of successive OFF hours, u is energy storage discharging state (1 if discharging, 0 otherwise), v is energy storage charging state (1 if charging, 0 otherwise), and z is adjustable load state (1 if on, 0 otherwise). The sum of power generated by DERs (i.e. dispatchable and non-dispatchable units and energy storage), power from the main grid, and the load curtailment must match the hourly load (4.2). Hourly load comprises consumption by both communal and individual loads. In grid-connected mode, LS will be zero, while in the islanded mode, P_M will be zero. The power transfer with the main grid is limited by the flow limits of the line connecting the microgrid to the main grid (4.3). Islanding is included in (4.3) by adding a binary islanding indicator, U_{ts} (the binary indicator is 0 when in islanded mode and 1 otherwise). The dispatchable unit generation is subject to minimum and maximum generation capacity limits (4.4), ramp-up and ramp-down rate limits (4.5) and (4.6), and minimum uptime and downtime limits (4.7) and (4.8). The change in the dispatchable unit generation is restricted by the generation adjustment limit when switched from the grid-connected to the islanded mode (4.9). The energy storage system power can be positive (discharging), negative (charging), or zero (idle) and is subject to charging and discharging minimum and maximum limits (4.10) and (4.12). The energy storage's available power is calculated based on the amount of charged/discharged power considering efficiency and restricted with capacity limits (4.13) and (4.14). The energy storage is subject to minimum charging and discharging time limits, i.e. the minimum number of consecutive hours that the energy storage should maintain its operating mode once it has changed (4.15) and (4.16). The adjustable loads are subject to minimum and maximum rated powers (4.17) and would consume the required energy to complete an operating cycle in the time intervals specified by the consumers (4.18). Loads could also be subject to minimum operating time, i.e. the number of consecutive hours that a load should consume power once it is switched on (4.19). In the proposed formulation, individual building loads cannot be controlled by the master controller; hence, (4.17)–(4.19) for every $d \in D_B$ must be solved in a different problem, which is managed by the local controller.

4.3.2 Local Controller

Local controller solves an optimal scheduling problem with the primary objective of minimizing the cost of supplying local loads (4.20) subject to constraints (4.21)–(4.23), as follows:

$$\text{Min} \sum_{s} \sum_{t} \sum_{d \in D_B} \pi_{ts} D_{dts} \tag{4.20}$$

s.t.

$$D_{dt}^{\min} z_{dt} \leq D_{dts} \leq D_{dt}^{\max} z_{dt} \quad \forall d \in D_B, \forall t, \forall s \tag{4.21}$$

$$\sum_{t \in [\alpha_d, \beta_d]} D_{dts} = E_d \quad \forall d \in D_B, \forall s \tag{4.22}$$

$$T_d^{\text{on}} \geq MU_d \left(z_{dt} - z_{d(t-1)} \right) \quad \forall d \in D_B, \forall t, s = 0 \tag{4.23}$$

The objective is subject to adjustable load constraints, including minimum and maximum rated powers (4.21), required energy to complete an operating cycle in the time intervals specified by consumers (4.22), and minimum operating time, i.e. the number of consecutive hours that a load should consume power once it is switched on (4.23). Buildings may further include local DERs, which could be directly considered in the objective, along with associated constraints.

Once optimal load schedule is calculated, the building net load, i.e. local load minus local generation, will be sent to the master controller. The challenge in this problem, however, is how to define proper prices for buildings and adjust the loads in a way that main grid power transfer limits are not violated.

4.3.3 Price Signal Calculation

Consider the master controller problem. The only variable that the master controller cannot control is the building load. Since this variable cannot be controlled, the power transfer with the main grid cannot be directly adjusted, and thus, there is a possibility to violate main grid power transfer limits imposed by (4.3). To resolve this issue, main grid power transfer limits are penalized with constant parameters, i.e. Lagrangian multipliers, in the objective function (4.1), where added costs are employed rather than strict inequality constraints, as follows:

$$\text{Min} \sum_{s} \sum_{t} \sum_{i \in G} c_{is} P_{its} + \sum_{s} \sum_{t} \rho_t P_{M,ts} + \sum_{s} \sum_{t} \upsilon_t LS_{ts}$$
$$+ \sum_{s} \sum_{t} \overline{\lambda}_{ts} \left(P_{M,ts} - P_M^{\max} U_{ts} \right) + \sum_{s} \sum_{t} \underline{\lambda}_{ts} \left(-P_M^{\max} U_{ts} - P_{M,ts} \right)$$

$$\tag{4.24}$$

where λ is a Lagrangian multiplier, and ρ is market price. The power transfer with the main grid, $P_{M,ts}$, can be replaced with its equivalent microgrid net load, obtained from (4.2). By substituting this variable and rearranging the terms, the objective function will be, as follows:

$$\text{Min} \sum_s \sum_t \sum_{i \in G} (c_{is} - \pi_{ts}) P_{its} + \sum_s \sum_t \sum_{d \in D_C} \pi_{ts} D_{dts} + \sum_s \sum_t \sum_{d \in D_B} \pi_{ts} D_{dts}$$
$$+ \sum_s \sum_t (v_t - \pi_{ts}) LS_{ts} - \sum_s \sum_t (\overline{\lambda}_{ts} + \underline{\lambda}_{ts}) P_M^{\max} U_{ts}$$

$$(4.25)$$

where

$$\pi_{ts} = \rho_t + \overline{\lambda}_{ts} - \underline{\lambda}_{ts} \qquad (4.26)$$

where π_{ts} is the revised price. This objective function includes revised prices for dispatchable generation units, adjustable loads, and load curtailment. The last term, as well as the term containing building loads, are constant in each iteration of the master controller problem and hence can be dropped from the objective. The obtained prices represent revised price signals, which will be used for scheduling DERs and loads. The revised market price, π_{ts}, is the price signal that will be sent to local controllers for load scheduling purposes as proposed in (4.20)–(4.23). The master controller problem could be accordingly rewritten as follows:

$$\text{Min} \sum_s \sum_t \sum_{i \in G} (c_{is} - \pi_{ts}) P_{its} + \sum_s \sum_t \sum_{d \in D_C} \pi_{ts} D_{dts} + \sum_s \sum_t (v_t - \pi_{ts}) LS_{ts},$$

$$(4.27)$$

subject to constraints (4.4)–(4.19).

The revised master controller problem only deals with directly controllable variables. The master controller starts with a zero value for penalty coefficients, schedules communal DERs and loads, and sends forecasted market prices to local controllers for load scheduling. Once schedules are determined and the net load data is gathered by the master controller, the main grid power transfer is determined for each scenario using Eq. (4.2). If violating any of the limits in (4.3), the associated Lagrangian multiplier will be updated for revising the solution in the next iteration. Lagrangian multipliers are updated as follows:

$$\overline{\lambda}_{ts}^{n+1} = \max\{0, \overline{\lambda}_{ts}^n + \overline{\sigma}(P_{M,ts} - P_M^{\max} U_{ts})\} \quad \forall t, \forall s \qquad (4.28)$$

$$\underline{\lambda}_{ts}^{n+1} = \max\{0, \underline{\lambda}_{ts}^n + \underline{\sigma}(-P_M^{\max} U_{ts} - P_{M,ts})\} \quad \forall t, \forall s \qquad (4.29)$$

where σ is the step size for Lagrangian multiplier update, and n is an index for iterations. Iterations will continue until the main grid power transfer limits are satisfied. The proposed formulation solves the microgrid optimal scheduling problem for both grid-connected and islanded modes. When the binary islanding indicator is set to zero, the main grid power transfer will be zero, and therefore, the microgrid is imposed to operate in the islanded mode. The binary islanding indicator will be further used to generate islanding scenarios. In each scenario, the binary islanding indicator will obtain a value of 0 or 1 based on the islanding duration and will be considered in the islanding scenarios as an input. Without loss of generality, an hourly scheduling is performed for the microgrid, in which islanding duration is also as an integer multiple of one hour. Shorter time periods can also be considered for embracing more data and increasing solution accuracy, which, however, would increase computation time.

4.3.4 Uncertainty Consideration

Uncertainty represents factors, which are out of control of the master and local controllers and/or cannot be determined with certainty. Prevailing uncertainties in this problem include forecast errors in non-dispatchable units' generation, loads, and market prices, as well as time and duration of islanding incidents. Uncertainty in islanding hours, including uncertain time and duration, can be efficiently modeled by defining islanding scenarios and specifying 0 and 1 values to the islanding indicator U_{ts}. The uncertainty in non-dispatchable units' generation can be ignored without a significant impact on results. As studied previously [21], the impact of non-dispatchable units' generation on microgrid optimal scheduling results is marginal, as these resources represent a small percentage of the generation mix, and, furthermore, are volatile and intermittent. Forecast errors can be directly included in the problem by defining forecasting scenarios. This action requires prior knowledge of the probability distribution functions of forecasted values. The large number of scenarios generated by Monte Carlo simulation can be further reduced using scenario reduction techniques to cope with computation burdens. Scenarios have already been included in the developed formulation; hence application of scenario-based stochastic programming to the proposed model would be simple and straightforward.

4.4 Case Study

A community microgrid test system is used to demonstrate the effectiveness of the proposed optimal scheduling model. The microgrid has nine communal DERs, including six dispatchable DERs with a total capacity of 24 MW, two non-dispatchable DERs with a total capacity of 2.5 MW, and one energy storage system with a rated charging/discharging power of 2 MW. The communal DERs characteristics are given in Tables 4.1 and 4.2. Three buildings are considered, each including a local controller. It is assumed that none of the buildings employ DERs; however, they utilize adjustable loads as given in Table 4.3. Aggregated adjustable loads are represented, in which it is assumed that each building can be modeled with one aggregated shiftable load and one aggregated curtailable load. Microgrid

Table 4.1 Characteristics of generating units.

Unit	Type	Marginal cost ($/MWh)	Min–max capacity (MW)	Min up/down time (h)	Ramp up/down rate (MW/h)
G1	D	27.7	1–5	3	2.5
G2	D	39.1	1–5	3	2.5
G3	D	45.4	1–5	3	2.5
G4	D	58.9	0.8–3	1	3
G5	D	61.3	0.8–3	1	3
G6	D	65.6	0.8–3	1	3
G7	ND	0	0–1	—	—
G8	ND	0	0–1.5	—	—

D, dispatchable; ND, non-dispatchable.

Table 4.2 Characteristics of the energy storage system.

Storage	Capacity (MWh)	Min–max charging/discharging power (MW)	Min charging/discharging time (h)
ESS	10	0.4–2	5

Table 4.3 Characteristics of adjustable loads.

Load	Building	Type	Min–max capacity (MW)	Required energy (MWh)
L1	1	S	0–4	16
L2	1	C	1.2–1.5	32
L3	2	S	0–4	24
L4	2	C	1.8–2.1	48
L5	3	S	0–6	40
L6	3	C	3–3.6	80

S, shiftable; C, curtailable.

aggregated fixed load, which is obtained from a day-ahead forecast, is given in Table 4.4. It is assumed that the fixed load distribution among buildings 1, 2, and 3 is 28%, 28%, and 44%, respectively. Tables 4.5 and 4.6 represent the generation of non-dispatchable units and the forecasted hourly market prices, respectively. The problem is implemented on a 2.4-GHz personal computer using CPLEX 11.0 [29].

Two cases are considered. The first case deals with the microgrid grid-connected operation to show how price signals would change the microgrid DER and load schedule to meet the prevailing limitations. The second case takes islanding into consideration.

Table 4.4 Microgrid hourly fixed load.

Time (h)	1	2	3	4	5	6
Load (MW)	12.96	12.67	12.57	13.41	13.05	13.08
Time (h)	7	8	9	10	11	12
Load (MW)	15.02	16.23	16.61	17.48	17.93	18.00
Time (h)	13	14	15	16	17	18
Load (MW)	20.66	22.66	22.80	23.30	23.95	23.99
Time (h)	19	20	21	22	23	24
Load (MW)	23.10	23.02	20.78	19.34	14.57	14.02

Table 4.5 Generation of non-dispatchable units.

Time (h)	1	2	3	4	5	6
G7	0	0	0	0	0.63	0.80
G8	0	0	0	0	0	0
Time (h)	7	8	9	10	11	12
G7	0.62	0.71	0.68	0.35	0.62	0.36
G8	0	0	0	0	0	0.75
Time (h)	13	14	15	16	17	18
G7	0.40	0.37	0	0	0.05	0.04
G8	0.81	1.20	1.23	1.28	1.00	0.78
Time (h)	19	20	21	22	23	24
G7	0	0	0.57	0.6	0	0
G8	0.71	0.92	0	0	0	0

Table 4.6 Hourly market price.

Time (h)	1	2	3	4	5	6
Price ($/MWh)	15.03	10.97	13.51	15.36	18.51	21.80
Time (h)	7	8	9	10	11	12
Price ($/MWh)	22.30	22.83	21.84	27.09	37.06	68.95
Time (h)	13	14	15	16	17	18
Price ($/MWh)	65.79	66.57	65.44	79.79	115.5	110.3
Time (h)	19	20	21	22	23	24
Price ($/MWh)	96.05	90.53	67.38	40.95	29.42	26.68

Case 1 (grid-connected mode): A main grid power transfer limit of 30 MW is considered. The microgrid optimal scheduling is performed, where first, the communal DERs are scheduled, followed by optimal load scheduling of each building. The initial solution shows a violation in calculated power transfer during hours 1–4, since the main grid price is low and loads are mainly shifted to these hours.

To eliminate the violation of these constraints, using a step of 0.01, the Lagrangian multipliers are calculated, and accordingly, the prices during hours 1–4 are updated. The final solution is obtained after 24 iterations with a computation time of 29 seconds. The generation of unit 1 is increased, and unit 2 is additionally committed. Moreover, shiftable loads are shifted away from hour 2 as it shows the largest violation. The total scheduling cost is $23,492.37, including $17,965.35 for local generation and $5527.02 for power transfer from the main grid.

To further analyze the microgrid optimal scheduling in response to power transfer limitations in grid-connected mode, the problem is solved for a tighter limit of 25 MW. Again, the initial solution violates the power transfer limit in hours 1–4. To eliminate violations in these hours, unit 1 generation is increased to its maximum limit, and adjustable loads are partially shifted to hours 5–9. This shift in the load results in a violation in hours 5–7. To resolve this, the price at these hours increases, which results in shifting back some of the adjustable loads to early morning hours and, accordingly, commitment of additional units 2 and 3. The final solution is obtained in 27 iterations with a computation time of 33 seconds and a total operation cost of $25,048.12.

The results show that a tighter limit on main grid power transfer causes an increase in the microgrid operations cost since the microgrid would be restricted in purchasing low-cost power from the main grid. However, it does not apply to cases where the purchased power from the main grid does not exceed its limits. In addition, the results from the latter case suggest that shifting loads to other operating hours to remove violations may result in temporary peak loads. These peak loads need to be further analyzed and eliminated to guarantee that power transfer limits are not violated, and if violated, additional rescheduling needs to be performed to address the issue.

Case 2 (Islanding consideration): Microgrid optimal scheduling problem is solved considering one islanding scenario with a single-hour islanding in hour 10 of the scheduling horizon. The final solution is obtained in 42 iterations (with a computation time of 54 seconds) with a total operations cost of $23,968.57. The cost of local generation and the cost of power transfer from the main grid are calculated as $18,796.4 and $5172.12, respectively. In this case, prices are updated in each iteration not only to obtain new schedules for DERs and loads in the grid-connected mode and satisfy main grid power transfer limits but also to reach a main grid power transfer value of zero in the islanding scenario. To ensure a seamless islanding, the grid-connected solution also needs to be modified, which will be efficiently performed by the proposed model. In the previous case, only units 1–3 are committed at hour 10 to supply a load of more than 25 MW, while the remaining required power is scheduled to be supplied from the main grid. In this case, additional units 4–6 are committed to supply a higher portion of the load. This increase in the available generation ensures that the microgrid would seamlessly switch to the islanded mode at hour 10, once the main grid power transfer is interrupted. The primary reason for the reduction in the cost of main grid power transfer and increase in the cost of local generation, compared to Case 1, is this change in the DER schedule. The result advocates that when considering islanding incidents in the microgrid, the grid-connected solution also needs to be revised to provide the required online capacity for a quick change in operation mode.

Studied cases require several iterations to reach the optimal solution. The main reasons for the large number of iterations are the step size used to update price signals as well as oscillations around the optimal point. If a larger step size is considered, violations will be reduced faster, however, it is possible that the new solutions oscillate around the optimal point, and hence, need further iterations. This issue is more critical in considering islanding incidents in which the microgrid net power should be exactly zero. Available solutions to these challenges include using more advanced Lagrangian relaxation methods and updating Lagrangian multipliers in a more efficient way. Extensive studies on improving the Lagrangian relaxation method are available in the literature, which will make this task easy. It is also noteworthy that the price curve of dispatchable generation units plays an important role in oscillations around the optimal point. When using single-step price curves, each step will be fully dispatched once the updated price is higher than the step price, which will accordingly result in a jump in microgrid generation. This is also true for the case when the updated price is lower than the step price, and hence there is a jump in generation reduction. To resolve this issue, a multi-step price curve with a large number of steps can be used. The appropriate number of steps can be found by addressing the tradeoff between the reduced number of iterations and added complexity due to increased problem variables.

Community microgrids introduce a novel and viable application of microgrids, which, however, introduce additional challenges to optimal scheduling of available resources due to customer privacy issues. Specific features of the proposed master controller design for the economic operations of community microgrid are, as follows:

- *Lower operations cost:* The proposed model determines the optimal schedule of DERs and loads, both communal and individual, along with the main grid power transfer, to minimize the total microgrid operation cost in grid-connected mode. Solution optimality in islanded modes can be simply adjusted using generation marginal costs.
- *Improved customer privacy:* The proposed model efficiently addresses the challenge of scheduling DERs and loads of microgrid buildings in which direct control is not acceptable due to customer privacy issues. To resolve this issue, price signals are developed by the master controller and used as control signals to revise building loads.
- *Islanding capability:* The microgrid optimal scheduling is reinforced with islanded operation constraints to provide sufficient capacity for a smooth and uninterrupted supply of loads when switching to an islanded mode. The proposed binary islanding indicator enables the utilization of deterministic and stochastic islanding scenarios for the problem.
- *Solution optimality:* The price signals are obtained using Lagrangian relaxation, and accordingly, the scheduling problem is decomposed into problems associated with the master controller and local controllers. It can be proven that the iterative problem will reach to optimal solution. Solution convergence can be further guaranteed using significant advances in Lagrangian relaxation methods.
- *Model scalability:* The proposed model can be applied to any microgrid size, independent of the number of buildings, DERs, and loads. It is also applicable to different types of microgrids without loss of generality.
- *Uncertainty consideration:* The proposed scenarios, which are primarily integrated into the model for incorporating islanding incidents, can be used to model uncertainties due to forecast errors associated with non-dispatchable generation, load, and market price forecasts. To do this, stochastic programming can be employed, in which a large number of scenarios are generated using Monte Carlo simulation and reduced to a computationally efficient number using a variety of scenario reduction techniques.

4.5 Summary

Community microgrids offer a viable alternative to centralized generation and distribution of electricity, while providing significant benefits to electricity customers and electric utilities. The design of an efficient master controller to be able to optimally schedule available DERs and loads while maintaining the privacy of customers, however, is of utmost importance for the economics of operations in these systems. An efficient model for community microgrid optimal scheduling considering customer privacy issues is proposed in this chapter. The objective of the proposed economic operations model is to minimize the microgrid's total operations cost, which comprises the generation cost of local resources and the cost of energy purchases from the main grid for supplying local loads. Microgrid DERs and loads are separated into two categories: communal and individual. Mathematically, it was formulated as a bi-level model. The upper level is solved by the microgrid master controller to schedule communal DERs and loads and further calculate price signals for local controllers. The lower level is solved by local controllers to schedule individual DERs and loads, and accordingly determine the net load of each building. The Lagrangian relaxation method was employed to decouple the master controller and local controllers scheduling problems. Price signals, which are formed in the master controller and sent to local controllers, and load data, which are calculated in local controllers and sent to the master controller, coupled the two problems. Hourly prices determined in the master controller and utilized in local controllers are the only available signals used by the master controller to control individual DERs and loads. The proposed model ensures the privacy of individual DERs and loads within the microgrid, prevents direct control by the master controller, finds the microgrid optimal schedule, and efficiently handles islanding incidents. The proposed model was analyzed through numerical simulations, where it was shown that the optimal microgrid schedule could be obtained using iterations among the master controller and local controllers while maintaining individual customers' privacy. Numerical simulations demonstrate the effectiveness of the proposed community microgrid optimal operations management model.

References

1 Electric Power Research Institute, "Assessment of Achievable Potential from Energy Efficiency and Demand Response Programs in the U.S," Electric Power Research Institute, 2009.

2 A. Ipackchi and F. Albuyeh, "Grid of the future—are we ready to transition to a smart grid?," *IEEE Power Energ. Mag.*, vol. 7, no. 2, pp. 52–62, Mar./Apr. 2009.

3 M. Shahidehpour, "Role of smart microgrid in a perfect power system," in *IEEE Power and Energy Society Gen. Meeting*, Minneapolis, MN, 2010.

4 A. Flueck and Z. Li, "Destination perfection," *IEEE Power Energ. Mag.*, vol. 6, no. 6, pp. 36–47, Nov./Dec. 2008.

5 M. Shahidehpour and J. F. Clair, "A functional microgrid for enhancing reliability, sustainability, and energy efficiency," *Electr. J.*, 25, no. 8, 21–28, Oct. 2012.

6 S. Bahramirad, W. Reder, and A. Khodaei, "Reliability-constrained optimal sizing of energy storage system in a microgrid," *IEEE Trans. Smart Grid*, vol. 3, no. 4, pp. 2056–2062, Dec. 2012.

7 N. Hatziargyriou, H. Asano, M. R. Iravani, and C. Marnay, "An overview of ongoing research, development and demonstration projects," *IEEE Power Energ. Mag.*, vol. 5, no. 4, pp. 79–94, July/Aug. 2007.

8 B. Kroposki, R. Lasseter, T. Ise, S. Morozumi, S. Papathanassiou, and N. Hatziargyriou, "Making microgrids work," *IEEE Power Energ. Mag.*, vol. 6, no. 3, pp. 40–53, May 2008.

9 I. Bae and J. Kim, "Reliability evaluation of customers in a microgrid," *IEEE Trans. Power Syst.*, vol. 23, no. 3, pp. 1416–1422, Aug. 2008.

10 S. Kennedy and M. Marden, "Reliability of Islanded Microgrids with Stochastic Generation and Prioritized Load," Bucharest: IEEE Powertech, June 2009.

11 F. Katiraei and M. R. Iravani, "Power management strategies for a microgrid with multiple distributed generation units," *IEEE Trans. Power Syst.*, vol. 21, no. 4, pp. 1821–1831, Nov. 2006.

12 A. Khodaei, "Resiliency-oriented microgrid optimal scheduling," *IEEE Trans. Smart Grid*, vol. 5, no. 4, pp. 1584–1591, July 2014.

13 A. Khodaei and M. Shahidehpour, "Microgrid-based co-optimization of generation and transmission planning in power systems," *IEEE Trans. Power Syst.*, vol. 28, no. 2, pp. 1582–1590, May 2013.

14 S. Chowdhury, S. P. Chowdhury, and P. Crossley, "Microgrids and Active Distribution Networks," IET Renewable Energy Series, 2009.

15 D. E. Olivares, C. A. Canizares, and M. Kazerani, "A centralized optimal energy management system for microgrids," in *Power Energy Soc. Gen. Meeting*, Detroit, MI, July 2011.

16 A. G. Tsikalakis and N. D. Hatziargyriou, "Centralized control for optimizing microgrids operation," *IEEE Trans. Energy Convers.*, vol. 23, no. 1, pp. 241–248, Mar. 2008.

17 N. Hatziargyriou, G. Contaxis, M. Matos, J. A. P. Lopes, G. Kariniotakis, D. Mayer, J. Halliday, G. Dutton, P. Dokopoulos, A. Bakirtzis, J. Stefanakis, A. Gigantidou, P. O'Donnell, D. McCoy, M. J. Fernandes, J. M. S. Cotrim, and A. P. Figueira, "Energy management and control of island power systems with increased penetration from renewable sources," in *IEEE-PES Winter Meeting*, New York, NY, vol. 1, pp. 335–339, Jan. 2002.

18 M. Korpas and A. T. Holen, "Operation planning of hydrogen storage connected to wind power operating in a power market," *IEEE Trans. Energy Conv.*, vol. 21, no. 3, pp. 742–749, Sept. 2006.

19 S. Chakraborty and M. G. Simoes, "PV-microgrid operational cost minimization by neural forecasting and heuristic optimization," in *IEEE Ind. App. Soc. Annual Meeting*, Edmonton, AB, Oct. 2008.

20 H. Vahedi, R. Noroozian, and S. H. Hosseini, "Optimal management of MicroGrid using differential evolution approach," in *7th International Conference on the European Energy Market (EEM)*, Madrid, Spain, June 2010.

21 A. Khodaei, "Microgrid optimal scheduling with multi-period islanding constraints," *IEEE Trans. Power Syst.*, vol. 29, no. 3, pp. 1383–1392, May 2014.

22 N. D. Hatziargyriou, A. Dimeas, A. G. Tsikalakis, J. A. P. Lopes, G. Karniotakis, and J. Oyarzabal, "Management of microgrids in market environment," in *International Conference on Future Power Systems*, Amsterdam, Netherlands, Nov. 2005.

23 T. Logenthiran, D. Srinivasan, and D. Wong, "Multi-agent coordination for DER in MicroGrid," in *IEEE International Conference on Sustainable Energy Technologies (ICSET)*, Singapore, Nov. 2008.

24 J. Oyarzabal, J. Jimeno, J. Ruela, A. Engler, and C. Hardt, "Agent based microgrid management system," in *International Conference on Future Power Systems*, Amsterdam, Netherlands, Nov. 2005.

25 A. Seon-Ju and M. Seung-Il, "Economic scheduling of distributed generators in a microgrid considering various constraints," in *Power Energy Soc. Gen. Meeting*, Calgary, AB, Canada, July 2009.

26 C. Gouveia, J. Moreira, C. L. Moreira, and J. A. Pecas Lopes, "Coordinating storage and demand response for microgrid emergency operation," *IEEE Trans. Smart Grid*, vol. 4, pp. 1898–1908, 2013.

27 J. Mitra and M. R. Vallem, "Determination of storage required to meet reliability guarantees on island-capable microgrids with intermittent sources," *IEEE Trans. Power Syst.*, vol. 27, no. 4, pp. 2360–2367, Nov. 2012.

28 R. Pawelek, I. Wasiak, P. Gburczyk, and R. Mienski, "Study on operation of energy storage in electrical power microgrid—modeling and simulation," in *International Conference on Harmonics and Quality of Power (ICHQP)*, Bergamo, Italy, Sept. 2010.

29 ILOG CPLEX, "ILOGs CsPLEX Homepage," [Online]. Available: www.ilog. com, 2009.

5

Provisional Microgrids for Renewable Energy Integration

5.1 Economic Considerations in Provisional Microgrids

Distributed energy resources (DERs) consist of distributed generators and energy storage, which can be installed at electricity consumers' premises to provide a local supply of loads. Based on this definition, DER deployments must have three distinct characteristics to be considered as a microgrid: (i) the electrical boundaries are clearly defined; (ii) a master controller is present to control and operate available resources as a single controllable entity; and (iii) the installed generation capacity exceeds the peak load for enabling islanded operation. Considering these characteristics, microgrids can be identified as small-scale power systems with the ability of self-supply and islanding, which could generate, distribute, and regulate the flow of electricity to local consumers.

Islanding is the most critical feature of microgrids. The microgrid islanding capability enables the microgrid to be disconnected from the main grid in case of upstream disturbances or voltage fluctuations [1–5]. Islanded operations of microgrids provide significant cost savings and load point reliability enhancements during major outages, which would economically justify the islanding design as part of microgrid planning decisions. This feature, however, may lead to some drawbacks as more microgrids are deployed around the world. First, islanded operations require that the microgrid's installed generation capacity exceed critical local loads. The microgrid master controller cannot rely on the generation of non-dispatchable units for this purpose. These units, which primarily include renewable energy resources such as solar and wind, produce a variable generation that cannot be controlled, and there is always a possibility that the forecasted generation is not materialized. This issue is also applicable to energy storage to a large extent, as islanded operation may occur when the energy storage is fully discharged. Therefore, microgrid developers commonly deploy a high percentage

The Economics of Microgrids, First Edition. Amin Khodaei and Ali Arabnya.
© 2024 The Institute of Electrical and Electronics Engineers, Inc.
Published 2024 by John Wiley & Sons, Inc.

of dispatchable energy resources, primarily in the form of gas-fired plants, and reduce the capacity of renewable resources to ensure reliable and seamless islanding at all times. This is especially the case when a higher level of initial capital expenditure for renewable energy resources is required compared to gas-fired plants. The second issue is the underutilization of the installed dispatchable capacity. The main grid power can benefit from the economies of scale in generation, and even after accounting for transmission and distribution costs and the associated losses, it is normally less expensive than the generation price of local dispatchable units. Local dispatchable units may be more economical than the main grid power when the transmission network is congested, and the real-time market price is high. This case, however, mainly occurs during the peak hours. The relatively short duration of the peak hours compared to the periods when the transmission network is not congested advocates that the microgrid would rely more on the main grid power rather than the locally generated power for supplying local loads. This issue would significantly impact the economic benefits from microgrid deployments, increase the anticipated return on investment, and economically disincentivize the deployment of this technology. The third issue, which is more critical for community microgrids in urban settings, is the placement of dispatchable units. Some DERs require a small space for installation, such as solar panels, which can be installed on consumers' rooftops. On the contrary, installation of dispatchable units in a neighborhood is not easy, and the required physical space, considering the necessary right of way, is not always freely available.

Early studies on microgrids can be found in [6–8], which were further followed by studies on various aspects of microgrids with a focus on economics, operations and control, role of power electronics, protection, and communication [9–11]. Microgrids introduce unique opportunities in power system operation and planning, such as improved reliability by introducing self-healing at the local distribution network and lowering the possibility of load shedding, higher power quality by managing local loads, reduction in carbon emissions by the diversification of energy sources, offering energy efficiency by responding to real-time market prices, and reducing the total system expansion planning cost by deferring investments on new generation and transmission facilities [12–20]. A discussion on existing microgrid optimal scheduling methods can be found in [4].

With the goal of addressing the economic and reliability needs of electricity consumers with less critical and sensitive loads, procuring distribution network flexibility offered by existing microgrids, and ensuring a rapid and widespread deployment of renewable energy resources in distribution networks, this chapter proposes the concept of *provisional microgrids*. The concept of provisional microgrids is built upon current studies on microgrids to make sure that similar benefits will be offered while the deployment of small-scale renewable energy resources is boosted. The optimal operations, comprising economic, reliability, and environmental merits of provisional microgrids, will be presented in this chapter as well.

Definition: Provisional microgrids are similar to microgrids as their electrical boundaries are clearly defined and a master controller operates available resources and controls the system. Unlike typical microgrids, however, provisional microgrids do not have the ability to be islanded on their own. Provisional microgrids, as the name suggests, are dependent on one or more electrically connected microgrids, namely *coupled microgrids* henceforth, for switching to an islanded mode. Provisional microgrids could utilize a high percentage of renewable energy resources without limitations on islanding requirements. When islanding is needed, the provisional microgrid would be disconnected from the main grid distribution network and rely on its own generation, as well as generation of the coupled microgrid, to supply local loads. Provisional microgrids can be considered as viable enabler for a more rapid deployment of variable-generation renewable energy resources in distribution networks and improvement of utilization rate of capital-intensive DERs in microgrids.

Rationale: The idea behind deployment of provisional microgrids is that by removing the islanding requirement, there is no need to deploy a high percentage of dispatchable units; hence, any generation mix can be deployed. Therefore, a high percentage of variable generation resources without concerning about islanding requirements can be installed. This deployment, however, is contingent upon low criticality and sensitivity of local loads. By deploying variable generation resources, it would be guaranteed that the installed capacity is not underutilized as the generation of these resources will be used once it is produced, regardless of energy price in the main grid. Moreover, connection to the coupled microgrid would provide the required flexibility to coordinate variable generation if needed, and the unused capacity of the coupled microgrid would be used in islanding incidents to ensure supply of local loads. To compare, the primary application of microgrids is to improve reliability for local customers and manage the ever-increasing penetration of DERs, while the primary application of provisional microgrids is to boost deployment of variable-generation renewable energy resources in distribution networks by leveraging the available flexibility offered by previously installed microgrids and providing economic and reliability benefits for local customers.

The idea of the provisional microgrid is different from that of interconnected microgrids. Interconnected microgrids, also known as microgrid clusters or super-microgrids, include two or more microgrids that are electrically connected and can exchange power among themselves in order to manage loads, reduce losses, and reduce energy purchases from the main grid [21]. In this setting, each microgrid can be individually disconnected from the main grid distribution network, as well as from other microgrids, to operate in the islanded mode. The provisional microgrid, however, does not have the capability to be islanded on its own.

Operations: The core operational structure of provisional microgrids is illustrated in Figure 5.1 and described as follows:

- Provisional microgrids generate energy by coordinating available resources and can interact with the main grid and the coupled microgrid for power transfer to supply local loads in normal (i.e. grid-connected) operational mode.
- Provisional microgrids disconnect from the main grid distribution network and transfer power with the coupled microgrid for supplying local loads in islanded operational mode.

It is assumed that the connection between the provisional microgrid and the coupled microgrid will be maintained during islanding operations. This connection will ensure mutual benefits for both coupled microgrid and provisional microgrid, since the coupled microgrid would benefit by selling its unused capacity to the provisional microgrid, and the provisional microgrid would purchase power in the islanded mode for increasing its reliability. The provisional microgrid would further rely on the coupled microgrid for frequency regulation and voltage control in case dispatchable unit deployment is limited in the provisional microgrid. Significant economic and reliability benefits stemming from the power transfer are momentous drivers in maintaining the connection between the provisional microgrid and the coupled microgrid in islanded modes. It is further assumed that the provisional microgrid and the coupled microgrid would operate simultaneously in the islanded mode in response to main grid disruptions and/or voltage fluctuations.

Provisional microgrids can deploy any selected DER generation mix to supply loads and maximize economic and reliability benefits without the requirement of fully supplying loads. In most of the operating hours, the power transfer with the main grid and the coupled microgrid helps supply local loads. In minor and

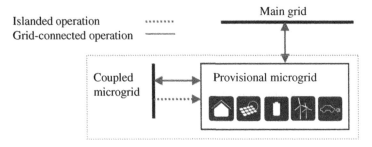

Figure 5.1 Core actions of the provisional microgrid.

infrequent islanding hours, the power transfer with the coupled microgrid combined with the adjustable load and energy storage schedules enables supplying local loads. The coupled microgrid is designed to completely supply its critical local loads at peak hours. Therefore, the coupled microgrid would normally have unused capacity in both grid-connected and islanded modes. The coupled microgrid's excess generation, beyond its load, would help the provisional microgrid to supply local loads during islanded operation. If sufficient generation is not available to fully supply loads, the provisional microgrid will curtail some of its load to guarantee load-supply balance. The possibility of load curtailment must be considered in the provisional microgrid design process as the cost of reliability.

5.2 An Economic Model for Provisional Microgrids

Figure 5.2 depicts the flowchart of the proposed optimal scheduling model. The problem is decomposed into a master problem and two subproblems. The master problem determines the optimal schedule of available dispatchable DERs as well

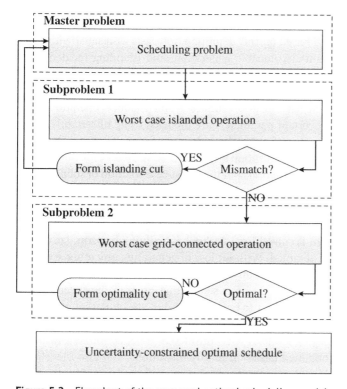

Figure 5.2 Flowchart of the proposed optimal scheduling model.

as adjustable loads. The obtained binary solution will be used in Subproblem 1, i.e. islanded operation, to examine power mismatches when islanded. If mismatches are not zero, i.e. sufficient generation is not available to supply local loads, the islanding cut is generated and added to the master problem for revising the current schedule. The islanding cut is represented in the form of an inequality constraint, which provides a lower estimate of the total mismatch in the subproblem as a function of scheduling variables in the master problem. The islanding-capable schedule, which is obtained in an iterative manner between the master problem and Subproblem 1, will be used in Subproblem 2, i.e. grid-connected operation, to determine the optimal dispatch of local DERs and the interactions with the main grid and the coupled microgrid. If the solution does not satisfy a predefined optimality criterion, the optimality cut is formed and sent back to the master problem for revising the current schedule. The optimality cut is represented in the form of an inequality constraint, which provides a lower estimate of the total operation cost as a function of scheduling variables in the master problem. The iterative procedure will continue until the final schedule, which meets both islanding and optimality criteria, is obtained.

The optimal scheduling problem is subject to several uncertainties. Uncertainty refers to the fact that some factors, having a major influence on scheduling decisions, are not under control of the microgrid master controller and/or cannot be predicted with certainty. Based on this definition, forecasts associated with fixed loads, market prices, non-dispatchable generation, and islanding incidents are considered as prevailing uncertainties in the scheduling process. Moreover, the information associated with the coupled microgrid is uncertain, which includes the available unused capacity in grid-connected and islanded operations and the generation price. Uncertain parameters are modeled using uncertainty intervals, which represent lower and upper bounds of deviations from nominal (i.e. forecasted) values. A robust optimization approach is adopted for capturing uncertainties. The robust optimization finds out the worst-case solution of subproblems as uncertain parameters vary within their associated uncertainty intervals. The robust optimization ensures that the obtained solution is robust against all realization of uncertain parameters [22–25].

The proposed problem is solved for a 24-hour scheduling horizon, i.e. a day-ahead schedule will be obtained. Without loss of generality, any other scheduling horizon can be selected based on the master controller's discretion in the proposed model. Selection of a 24-hour scheduling horizon, however, would enable microgrid master controller to benefit from day-ahead market price forecasts provided by the utility company and can keep track of the energy storage daily charging/discharging cycles. The considered time period is one hour, where schedules are obtained based on hourly operations, and also the islanding duration is considered as an integer multiple of one hour. Shorter time periods could be employed to capture rapid changes more accurately in load and non-dispatchable generation, as well as shorter islanding durations. The selection of a proper time period for scheduling represents a tradeoff between the solution accuracy and the computation time. Shorter time periods would analyze more data and provide more accurate solutions while increasing computation requirements.

5.2.1 Component Modeling

Dispatchable units, energy storage, and adjustable loads are scheduled by the microgrid master controller. A mixed-integer programming representation of these components is required as a primary step to model microgrid scheduling problem. The component models can be found in [4], however, are briefly presented here for further use in the problem formulation. Non-dispatchable generations as well as fixed loads are obtained based on forecasts, hence treated as constants in the problem formulation.

$$P_i^{\min} I_{it} \le P_{it} \le P_i^{\max} I_{it} \qquad \forall i \in G, \forall t \tag{5.1}$$

$$P_{it} - P_{i(t-1)} \le UR_i \qquad \forall i \in G, \forall t \tag{5.2}$$

$$P_{i(t-1)} - P_{it} \le DR_i \qquad \forall i \in G, \forall t \tag{5.3}$$

$$T_i^{\mathrm{on}} \ge UT_i\left(I_{it} - I_{i(t-1)}\right) \qquad \forall i \in G, \forall t \tag{5.4}$$

$$T_i^{\mathrm{off}} \ge DT_i\left(I_{i(t-1)} - I_{it}\right) \qquad \forall i \in G, \forall t \tag{5.5}$$

$$P_{it} \le P_{it}^{\mathrm{dch,\,max}} u_{it} - P_{it}^{\mathrm{ch,\,min}} v_{it} \qquad \forall i \in S, \forall t \tag{5.6}$$

$$P_{it} \ge P_{it}^{\mathrm{dch,\,min}} u_{it} - P_{it}^{\mathrm{ch,\,max}} v_{it} \qquad \forall i \in S, \forall t \tag{5.7}$$

$$C_{it} = C_{i(t-1)} - P_{it} u_{it} \tau / \eta_i - P_{it} v_{it} \tau \qquad \forall i \in S, \forall t \tag{5.8}$$

$$C_i^{\min} \le C_{it} \le C_i^{\max} \qquad \forall i \in S, \forall t \tag{5.9}$$

$$T_i^{\mathrm{ch}} \ge MC_i\left(u_{it} - u_{i(t-1)}\right) \qquad \forall i \in S, \forall t \tag{5.10}$$

$$T_i^{\mathrm{dch}} \ge MD_i\left(v_{it} - v_{i(t-1)}\right) \qquad \forall i \in S, \forall t \tag{5.11}$$

$$u_{it} + v_{it} \le 1 \qquad \forall i \in S, \forall t \tag{5.12}$$

$$D_{dt}^{\min} z_{dt} \le D_{dt} \le D_{dt}^{\max} z_{dt} \qquad \forall d \in D, \forall t \tag{5.13}$$

$$\sum_{t \in [\alpha_d, \beta_d]} D_{dt} = E_d \qquad \forall d \in D \tag{5.14}$$

$$T_d^{on} \geq MU_d\left(z_{dt} - z_{d(t-1)}\right) \qquad \forall i \in D, \forall t \tag{5.15}$$

where ch is superscript for energy storage system charging mode, d is index for loads, dch is superscript for energy storage system discharging mode, i is index for DERs, t is index for time, $\wedge$ is index for calculated variables, D is set of adjustable loads, G is set of dispatchable units, P is set of primal variables, S is set of energy storage systems, U is set of uncertain parameters, c is marginal cost of dispatchable units, DR is ramp-down rate, DT is the minimum downtime, E is load total required energy, MC is minimum charging time, MD is minimum discharging time, MU is minimum operating time, UR is ramp-up rate, UT is minimum uptime, C is energy storage available (stored) energy variable, D is load demand variable, I is commitment state variable of dispatchable unit (1 when committed, 0 otherwise), P is DER output power variable, Q is operation cost variable, T^{ch} is the variable for umber of successive charging hours, T^{dch} is the variable for the number of successive discharging hours, T^{on} is the variable for number of successive ON hours, T^{off} is the variable for number of successive OFF hours, τ is the variable for time period, u is the variable for energy storage discharging state (1 when discharging, 0 otherwise), v is the variable for energy storage charging state (1 when charging, 0 otherwise), and z is the variable for adjustable load state (1 when operating, 0 otherwise).

The dispatchable unit generation is subject to minimum and maximum generation capacity limits (5.1), ramp-up and ramp-down rate limits (5.2) and (5.3), and minimum up and downtime limits (5.4) and (5.5). A dispatchable unit can be further subject to fuel and emission limits based on the unit type.

The energy storage power is subject to charging and discharging minimum and maximum limits depending on its mode (5.6) and (5.7). The energy storage charging power is considered negative, so the associated limits are denoted with a minus sign. Energy storage available energy is calculated based on the amount of charged/discharged power and efficiency (5.8) and restricted with capacity limits (5.9). Hourly studies are performed where the time period is considered to be one hour, i.e. $\tau =$ one hour. The available energy at $t = 1$ is calculated based on the available energy at the last hour of the previous scheduling horizon. The energy storage is subject to minimum charging and minimum discharging time limits, respectively, (5.10) and (5.11), which are the minimum number of consecutive hours that the energy storage should maintain charging/discharging once it changes its operational mode. It is further ensured that the energy storage is operated at one of the charging and discharging modes every hour (5.12).

Adjustable loads are subject to minimum and maximum rated powers (5.13). Each load consumes the required energy to complete an operating cycle in the time intervals specified by consumers (5.14). α_d and β_d, respectively, represent the start and end operating times of an adjustable load. Certain loads may be subject to minimum operating time, which is the number of consecutive hours that a load should consume power once it is switched on (5.15). The proposed formulation is applicable to adjustable loads that could be curtailed (i.e. curtailable loads) or deferred (i.e. shiftable loads).

5.2.2 Problem Formulation

Master Problem – Scheduling: The master problem is formulated, as follows:

$$\min \sum_t \sum_{i \in G} [c_{i0} I_{it} + SU_i + SD_i] + \sum_{d \in D} K_d \Delta_d + \Lambda \tag{5.16}$$

Subject to (5.4) and (5.5), (5.10)–(5.12), and (5.15), where c_0 is no-load cost, K_d is inconvenience penalty factor, SU startup cost variable, SD is shutdown cost variable, Δ_d is the variable for deviation in adjustable load operating time interval, and Λ is the variable for reflected operations cost in the master problem. The master problem objective function (5.16) comprises three terms associated with dispatchable units, adjustable loads, and the grid-connected operation cost. The no-load, startup, and shutdown costs of dispatchable units are calculated in the master problem since all are dependent only on binary commitment variables. The operations costs of the energy storage and adjustable loads are zero. The inconvenience encountered by consumers to revise their adjustable load operating time interval is considered in the second term of the objective, where $\Delta_d = \left(\beta_d^{new} - \alpha_d^{new}\right) - (\beta_d - \alpha_d)$ is the total change in the operating time interval, where α, β denote specified start and end times of adjustable loads. The inconvenience cost is represented as a penalty term times the total change in the operating time interval. This term prioritizes adjustable loads based on their criticality to be operated at the specified time interval. The last term in (5.16) is the grid-connected operation cost, which is calculated in Subproblem 2 and reflected in the master problem via optimality cuts.

The proposed master problem formulation includes only binary scheduling variables associated with dispatchable units, energy storage, and adjustable loads. Clearly, this problem will result in an all-zero solution in the first iteration. This solution, however, will be revised in subsequent iterations as islanding and optimality cuts are generated in subproblems and added to the master problem for governing the master problem solution.

Subproblem 1 – Islanded Operations: The objective of the islanded operation problem is to minimize power mismatches when islanded (5.17).

$$\max_U \min_P w = \sum_t (SL_{1,t} + SL_{2,t}) \tag{5.17}$$

Subject to (5.1)–(5.3), (5.6)–(5.9), (5.13), and (5.14), and

$$\sum_i P_{it} + P_{M,t} + P_{CM,t} + SL_{1,t} - SL_{2,t} = \sum_d D_{dt} \qquad \forall t \tag{5.18}$$

$$-P_M^{\max} U_t \le P_{M,t} \le P_M^{\max} U_t \qquad \forall t \tag{5.19}$$

$$P_{CM,t}^{\min} \le P_{CM,t} \le P_{CM,t}^{\max} \qquad \forall t \tag{5.20}$$

where w is the variable for power mismatch, P_{CM} is coupled microgrid power variable, P_M is main grid power variable, SL_1, SL_2 are slack variables, and U is outage state of main grid line/islanding state (0 when islanded, 1 otherwise). The power balance Eq. (5.18) ensures that the sum of power generated by DERs, power from the main grid, and power from the coupled microgrid matches the hourly load. The energy storage power can be positive (discharging), negative (charging), or zero (idle). The main grid power can be positive (import), negative (export), or zero. The coupled microgrid power can be positive (import), negative (export), or zero. Slack variables, which are both nonnegative, characterize virtual generation and load in the provisional microgrid and represent the mismatch between the available generation and the load. The provisional microgrid power transfer with the main grid is limited by the flow limit of the associated connecting line (5.19). The provisional microgrid power transfer with the coupled microgrid is limited by the coupled microgrid's available unused capacity limits (5.20). The binary outage state U_t is included in the main grid power transfer constraint to model islanded operation. When the binary outage state is set to zero, the main grid power will be zero; hence, the provisional microgrid is enforced to operate in the islanded mode. The power transfer with the coupled microgrid, however, could always be nonzero. Islanding is considered as an uncertain parameter in this problem, thus the worst-case scenario solution associated with islanding incidents will be obtained. The number of islanding hours, moreover, will be restricted by a limit on uncertainty option as $\sum_t U_t \le U^{\max}$, where $U^{\max}$ is the maximum number of islanding hours in the scheduling horizon.

If the objective is not zero, i.e. sufficient generation is not available to supply local loads, the islanding cut is formed and added to the master problem for revising the current schedule. The islanding cut is defined as follows:

$$\begin{aligned}
&\hat{w} + \sum_t \sum_{i \in G} \lambda_{it} \left(I_{it} - \hat{I}_{it} \right) + \sum_t \sum_{i \in S} \mu_{it}^{dch} \left(u_{it} - \hat{u}_{it} \right) \\
&+ \sum_t \sum_{i \in S} \mu_{it}^{ch} \left(v_{it} - \hat{v}_{it} \right) + \sum_t \sum_{i \in D} \pi_{dt} \left(z_{dt} - \hat{z}_{dt} \right) \le 0
\end{aligned} \tag{5.21}$$

where λ_{it}, μ_{it}^{dch}, μ_{it}^{ch}, and π_{dt} are dual variables associated with dispatchable unit commitment states, energy storage discharging state, energy storage charging state, and adjustable load scheduling state, respectively. It is statistically possible that after a certain number of iterations and revising the master problem solution, a feasible islanding is not achieved and the power mismatch still persists. The microgrid master control will, therefore, curtail loads. This action is considered as the last resort since it causes a significant inconvenience for microgrid consumers. The microgrid master controller will simply curtail the load, equal to the power mismatch between the available generation and the load, to achieve a feasible islanding. Once curtailed, the obtained feasible schedule will be sent to the grid-connected operation problem.

Subproblem 2 – Grid-Connected Operations: The objective of the grid-connected operations problem is to minimize the microgrid's total operation cost (5.22).

$$\max_{U} \min_{P} \ Q = \sum_{t}\sum_{i \in G} c_i P_{it} + \sum_t \rho_{M,t} P_{M,t} + \sum_t \rho_{CM,t} P_{CM,t} \tag{5.22}$$

Subject to (5.1)–(5.3), (5.6)–(5.9), (5.13), and (5.14), and

$$\sum_i P_{it} + P_{M,t} + P_{CM,t} = \sum_d D_{dt} - LS_t \qquad \forall t \tag{5.23}$$

$$- P_M^{\max} \le P_{M,t} \le P_M^{\max} \qquad \forall t \tag{5.24}$$

$$P_{CM,t}^{\min} \le P_{CM,t} \le P_{CM,t}^{\max} \qquad \forall t \tag{5.25}$$

where LS is the load curtailment in islanded operation, ρ_{CM} is coupled microgrid generation price, and ρ_M is market price. The first term in the objective function (5.22) is the operations cost of dispatchable units in the provisional microgrid, which includes generation cost over the entire scheduling horizon. The no-load, startup, and shutdown costs are excluded, as these costs are already considered in the master problem objective. The generation cost is approximated by a single-step linear model. The second term is the cost of power transfer from the main grid based on the market price at the point of common coupling. When the provisional microgrid excess power is sold back to the main grid, $P_{M,t}$ would be negative, so this term would represent a benefit rather than a cost. The last term is the cost of power transfer from the coupled microgrid. When the provisional microgrid excess power is sold back to the coupled microgrid $P_{CM,t}$ would be negative, so this term would represent a benefit rather than a cost. The power balance Eq. (5.23) ensures that the sum of power generated by DERs, power from the main grid, and power from the coupled microgrid matches the revised load, i.e. the provisional microgrid hourly load minus the load curtailment calculated in islanded operation. The provisional microgrid power transfer with the main grid is limited by the flow limits of the associated connecting line (5.24). The provisional microgrid power transfer with the coupled microgrid is limited by the coupled microgrid's available unused capacity limits (5.25). Since the provisional microgrid is grid-connected, the binary outage state is not considered in (5.24).

The solution optimality is examined by comparing an upper bound (obtained from the grid-connected operation problem) and a lower bound (which is the solution of the master problem). The proximity of two bounds ensures solution optimality; otherwise, the optimality cut (5.26) is generated and added to the master problem for revising the current schedule.

$$
\Lambda \geq \hat{Q} + \sum_t \sum_{i \in G} \lambda_{it} \left(I_{it} - \hat{I}_{it} \right) + \sum_t \sum_{i \in S} \mu_{it}^{dch} \left(u_{it} - \hat{u}_{it} \right) + \sum_t \sum_{i \in S} \mu_{it}^{ch} \left(v_{it} - \hat{v}_{it} \right)
$$
$$
+ \sum_t \sum_{i \in D} \pi_{dt} \left(z_{dt} - \hat{z}_{dt} \right)
$$

(5.26)

where λ_{it}, μ_{it}^{dch}, μ_{it}^{ch}, and π_{dt} are dual variables associated with dispatchable unit commitment states, energy storage discharging state, energy storage charging state, and adjustable load state, respectively. $\hat{Q}$ is the calculated objective value of the grid-connected operation problem.

Using robust optimization, the worst-case scenario solution of subproblems will be achieved where the objectives are represented in the form of max–min optimization. Objectives are minimized over primal variables and maximized over uncertain parameters. To solve complex max–min optimization subproblems, the dual problem of the inner minimization problem is found in each subproblem and combined with the associated maximization problem. The problem is accordingly solved for the set of dual variables and uncertain parameters [24, 25]. The employed robust optimization captures uncertainties in load, non-dispatchable generation, market prices, and islanding, as well as available unused capacity and price of the coupled microgrid. The level of uncertainty related to each uncertain parameter could be further adjusted by adding a limit on uncertainty options [26]. The proposed uncertainty-constrained optimal scheduling model is used for a single provisional microgrid and will be solved by its respective master controller. The coupled microgrid revenue from the optimal scheduling problem will be equal to the cost of power transfer from the coupled microgrid, as represented in the last term of (5.22), and will be calculated for both grid-connected and islanded modes.

5.3 Case Study

A provisional microgrid test system with three non-dispatchable units, one energy storage, and five adjustable loads is used to analyze the proposed optimal scheduling model and investigate the provisional microgrid's economic operation.

The problem is implemented on a 2.4-GHz personal computer using CPLEX 11.0 [27]. The characteristics of adjustable loads are given in Table 5.1. The forecasted values for hourly fixed load, aggregated non-dispatchable generation, and market prices over a 24-hour planning horizon are, respectively, given in Tables 5.2–5.4, with respective forecast errors of $\pm 10\%$, $\pm 20\%$, and $\pm 20\%$. The capacity of the energy storage is 10 MWh with min–max charging/discharging power of 0.4–2 MW, respectively. A minimum charging/discharging time of five hours is considered for the energy storage, i.e. the minimum number of consecutive hours that the energy storage must maintain its current operational state once the operational mode is changed. The main grid power transfer limit is 10 MW.

The proposed case studies focus on the coupled microgrid's available unused capacity to investigate the behavior of the provisional microgrid and calculate costs and benefits. It is assumed that the provisional microgrid requirements for power import during islanded operation are considered in the design process,

Table 5.1 Characteristics of adjustable loads.

Load	Type	Min–max capacity (MW)	Required energy (MWh)	Initial start-end time (h)	Min up time (h)
L1	S	0–0.4	1.6	11–15	1
L2	S	0–0.4	1.6	15–19	1
L3	S	0.02–0.8	2.4	16–18	1
L4	S	0.02–0.8	2.4	14–22	1
L5	C	1.8–2	47	1–22	24

S, Shiftable; C, Curtailable.

Table 5.2 Microgrid hourly fixed load.

Time (h)	1	2	3	4	5	6
Load (MW)	1.86	1.82	1.81	1.92	1.88	1.88
Time (h)	7	8	9	10	11	12
Load (MW)	2.16	2.33	2.39	2.51	2.58	2.59
Time (h)	13	14	15	16	17	18
Load (MW)	2.97	3.26	3.28	3.35	3.44	3.44
Time (h)	19	20	21	22	23	24
Load (MW)	3.32	3.31	2.99	2.78	2.10	2.02

Table 5.3 Aggregated generation of non-dispatchable units.

Time (h)	1	2	3	4	5	6
Power (MW)	0	0	0	0	2.52	3.20
Time (h)	7	8	9	10	11	12
Power (MW)	2.48	2.84	2.72	2.40	2.48	4.44
Time (h)	13	14	15	16	17	18
Power (MW)	4.84	6.27	4.93	5.12	4.21	3.28
Time (h)	19	20	21	22	23	24
Power (MW)	2.84	3.68	2.29	2.40	0	0

Table 5.4 Hourly market price.

Time (h)	1	2	3	4	5	6
Price ($/MWh)	15.03	10.97	13.51	15.36	18.51	21.80
Time (h)	7	8	9	10	11	12
Price ($/MWh)	17.30	22.83	21.84	27.09	37.06	68.95
Time (h)	13	14	15	16	17	18
Price ($/MWh)	65.79	66.57	65.44	79.79	115.5	110.3
Time (h)	19	20	21	22	23	24
Price ($/MWh)	96.05	90.53	77.38	70.95	59.42	56.68

i.e. the coupled microgrid can provide the provisional microgrid with sufficient generation for ensuring seamless islanding. The coupled microgrid's maximum hourly unused capacity is assumed to be 4 MW. The amount of available unused capacity, however, is uncertain, which will be considered in the provisional microgrid scheduling problem.

5.3.1 Case 1: A Baseline Case with Load, Non-Dispatchable Generation, and Market Price Uncertainties

In this case, the uncertainty in the power transfer with the coupled microgrid is not considered, i.e. the provisional microgrid can import/export power from/to the coupled microgrid up to 4 MW at each scheduling hour in grid-connected and islanded modes. The worst-case islanding occurs at hour 4, when the provisional microgrid does not have any local generation. The provisional microgrid is disconnected from the main grid at this hour but is still connected to the coupled

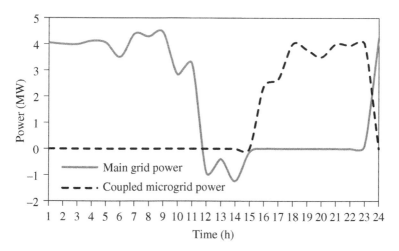

Figure 5.3 Power transfers with the main grid and the coupled microgrid.

microgrid. A power transfer of 3.92 MW from the coupled microgrid ensures seamless islanding where the load is fully supplied. The imported power supplies the fixed load, considering 10% uncertainty, and the curtailable load, L5. The provisional microgrid's total operations cost is $3066.50. Power transfers with the main grid and the coupled microgrid result in power purchase costs of $1088.53 and $1977.97, respectively, as shown in Figure 5.3. This figure exhibits that the provisional microgrid would consider the coupled microgrid as an alternative to purchasing power when the main grid power is more expensive. In addition, during hours 12–14, when the provisional microgrid generation exceeds its load, the excess generation is sold back to the main grid to increase economic benefits and reduce the operations costs. The charging/discharging schedule of the energy storage is adjusted based on the islanding hour, in which the energy storage charging is started at hour 5 to reduce the required power to be purchased from the coupled microgrid in islanded operation. The obtained results demonstrate how uncertainties, which are considered in subproblems, would impact the scheduling decisions in the master problem. Considering only provisional microgrid economics, the energy storage must be charged at hours 1–5; however, this schedule is revised in subsequent iterations as the worst-case islanding solution is identified in the islanded operations problem.

5.3.2 Case 2: Considering Uncertainty in the Coupled Microgrid's Available Unused Capacity

An uncertainty of 25% is considered in the coupled microgrid's available unused capacity in islanded and grid-connected modes. The grid-connected schedule will differ from Case 1, where the available grid-connected power from the coupled microgrid in hours 16–23 is reduced to 3 MW. The microgrid would reschedule adjustable loads by shifting away from these hours, as it is a more economical solution than purchasing relatively more expensive power from the main grid. The energy storage schedule is also changed as it is discharged for a longer period. The islanding, which occurs at hour 4, results in 0.92 MW load curtailment. The total cost, in this case, is $3069.70 with a cost breakdown of $1188.91 in power purchase from the main grid and $1880.79 in power purchase from the coupled microgrid.

Comparing Case 1 and Case 2, it can be concluded that the provisional microgrid can economically benefit from coupled microgrid grid-connected power and from islanding capability. The installed generation capacity in the provisional microgrid is fully utilized without any concern regarding capacity underutilization. The cost paid for preventing underutilization is infrequent load curtailments during islanding incidents. It is noteworthy that the load curtailment is obtained considering the worst-case solution under all uncertainties associated with loads, market prices, non-dispatchable generation, the coupled microgrid's available unused capacity, and the time of the islanding. Thus, the obtained value represents the maximum possible load curtailment under considered assumptions, and the actual value will be lower than this.

If loads are critical and require zero-load curtailment during islanding, the installed generation mix can be reinforced with dispatchable DERs. Two alternatives will be sought here to prevent the load curtailment in the islanded mode, as follows: (i) using faster energy storage with a lower minimum charging/discharging time; and (ii) using a relatively small dispatchable unit.

5.3.3   Case 2(a): Employing Fast Charge/Discharge Energy Storage

Using energy storage with a 2-hour minimum charging/discharging time not only improves the provisional microgrid's economic operation but also removes load curtailment in islanded operations. The total operations cost in this case is reduced to $3003.00, and the load curtailment in islanded operation at hour 4 is reduced down to zero. The energy storage is charged at hours 1–3 and discharged at hour 4 to provide the required power to supply local loads. The energy storage encounters three charging/discharging cycles in this case, compared to only one cycle in Case 1. This case study exhibits that the provisional microgrid could still rely on non-dispatchable units for grid-connected operation and benefit from fast charge/discharge energy storage to ensure seamless islanding.

5.3.4 Case 2(b): Addition of a 1 MW Dispatchable Unit

A 1 MW dispatchable unit is added to the provisional microgrid with a single-step generation price of $75/MWh. The addition of this unit will reduce the load curtailment to zero. Moreover, this unit will generate power during high-price hours for reducing the power purchase from the main grid. The total operations cost is $3047.70, including $567.87 generation cost of the dispatchable unit. Although dispatchable unit is added to the provisional microgrid, a high percentage of the installed capacity would be non-dispatchable. Thus, the provisional microgrid would mainly rely on non-dispatchable generation and energy storage for its economic and reliable operation.

These two cases suggest that provisional microgrids can represent a viable solution to economically supply local loads and achieve desired reliability targets while employing a high degree of non-dispatchable generation, primarily in the form of variable-generation renewable energy resources. Moreover, the emissions produced by the provisional microgrid are much lower than a microgrid of the same size, which mainly relies on gas-fired plants. Therefore, the provisional microgrid can significantly support environmental objectives and be considered as a sustainable option to enable large-scale deployment of renewable energy resources.

Although significant social cost savings and load point reliability enhancements offered by islanding justify the islanding design as part of the microgrid planning decisions, the obtained increased investment cost, underutilization of dispatchable units, and under-deployment of renewable energy resources necessitates introduction of new classes of microgrids. Provisional microgrids can address these challenges. Specific features of provisional microgrids and the proposed optimal scheduling model are, as follows:

- *Avoiding capacity underutilization*: Provisional microgrids reduce the need to build microgrids with high dispatchable generation capacity and the possibility of installed capacity underutilization. Non-dispatchable units constitute the majority of installed capacity in provisional microgrids. which will generate power independent of market price variations.
- *Removing the need to enhance distribution network flexibility*: Provisional microgrids will benefit from the available flexibility in distribution networks offered by existing microgrids. Thus, there would be no need for system operators to reinforce the distribution network's flexibility by additional installations and system upgrades.
- *Reduced tension on transmission and distribution networks*: Provisional microgrids improve power system operational efficiency in integrating local DERs. Consequently, the tension on congested transmission and distribution networks

will be reduced, which will benefit system developers by deferring required system upgrades.

- *Environmental impacts*: Provisional microgrids address environmental concerns by enabling large and distributed penetration of emission-free variable-generation renewable energy resources in distribution networks and helping to transition from large, centralized coal and gas-fired plants.

5.4 Summary

A new class of microgrids, namely provisional microgrids, was introduced in this chapter. Provisional microgrids share similar characteristics as typical microgrids; however, do not possess the islanding capability and are dependent on one or more electrically connected microgrids for islanding purposes. Removing the islanding requirements and relying on the available unused capacity of existing microgrids characterizes provisional microgrids as enablers of rapidly deploying variable-generation renewable energy resources in distribution networks and further reducing underutilization of capital-intensive DERs in microgrids. Provisional microgrids are defined, and an uncertainty-constrained optimal scheduling model is developed, which considers prevailing uncertainties associated with loads, non-dispatchable generation, and market price forecasts, as well as islanding incidents and the available unused capacity from coupled microgrids. An uncertainty-constrained optimal scheduling model was proposed to efficiently model the day-ahead operations of provisional microgrids considering prevailing operational uncertainties. Robust optimization was employed, where the original problem was decomposed into smaller and coordinated problems for uncertainty consideration. The optimal scheduling problem is decomposed using Benders' decomposition and solved via the robust optimization method. The proposed model was analyzed through numerical simulations, and it was shown that provisional microgrids offer economic benefits, ensure reliability, and prevent underutilization of deployed capital-intensive DERs.

Future studies on this topic include, but are not limited to: (i) optimal planning of provisional microgrids with the objective of economically justifying the provisional microgrid deployment; (ii) control studies for ensuring that frequency and voltages within the provisional microgrid could be efficiently controlled and maintained within limits during grid-connected and islanded modes; (iii) assessing the distribution network hosting capacity when integrating increased levels of non-dispatchable generation via provisional microgrids; and (iv) communication studies for ensuring that the information can be reliably exchanged between provisional microgrids and coupled microgrids.

References

1 M. Shahidehpour, "Role of smart microgrid in a perfect power system," in *IEEE Power and Energy Society Gen. Meeting*. Minneapolis, MN, 2010.

2 A. Flueck and Z. Li, "Destination perfection," *IEEE Power Energ. Mag.*, vol. 6, no. 6, pp. 36–47, 2008.

3 M. Shahidehpour and J. Clair, "A functional microgrid for enhancing reliability, sustainability, and energy efficiency," *Electr. J.*, vol. 25, pp. 21–28, 2012.

4 A. Khodaei, "Microgrid optimal scheduling with multi-period islanding constraints," *IEEE Trans. Power Syst.*, vol. 29, pp. 1383–1392. Accepted for publication, 2014.

5 S. Bahramirad, W. Reder, and A. Khodaei, "Reliability-constrained optimal sizing of energy storage system in a microgrid," *IEEE Trans. Smart Grid*, vol. 3, no. 4, pp. 2056–2062, 2012.

6 C. Marnay, F. Rubio, and A. Siddiqui, "Shape of the microgrid," in *IEEE Power Engineering Society Winter Meeting*, Columbus, OH, vol. 1, pp. 150–153, 2001.

7 A. Meliopoulos, "Challenges in simulation and design of µgrids," in *IEEE Power Engineering Society Winter Meeting*, New York, NY, vol. 1, pp. 309–314, 2002.

8 R. Lasseter, "Microgrids," in *IEEE Power Engineering Society Winter Meeting*, New York, NY, vol. 1, pp. 305–308, 2002.

9 S. Chowdhury, S. P. Chowdhury, and P. Crossley, "Microgrids and Active Distribution Networks," IET Renewable Energy Series, The Institution of Engineering and Technology, Athenaeum Press, 2009.

10 H. Jiayi, J. Chuanwen, and X. Rong, "A review on distributed energy resources and MicroGrid," *Renew. Sust. Energ. Rev.*, vol. 12, no. 9, pp. 2472–2483, 2008.

11 N. Hatziargyriou, "Microgrids: Architectures and Control," Wiley, 2014.

12 A. Khodaei, "Resiliency-oriented microgrid optimal scheduling," *IEEE Trans. Smart Grid*, vol. 5, pp. 1584–1591. Accepted for publication, 2014.

13 B. Kroposki, R. Lasseter, T. Ise, S. Morozumi, S. Papathanassiou, and N. Hatziargyriou, "Making microgrids work," *IEEE Power Energ. Mag.*, vol. 6, no. 3, pp. 40–53, 2008.

14 I. Bae and J. Kim, "Reliability evaluation of customers in a microgrid", *IEEE Trans. Power Syst.*, vol. 23, no. 3, pp. 1416–1422, 2008.

15 S. Kennedy and M. Marden, "Reliability of Islanded Microgrids with Stochastic Generation and Prioritized Load," Bucharest: IEEE Powertech, June 2009.

16 N. Hatziargyriou, H. Asano, M. R. Iravani, and C. Marnay, "Microgrids: an overview of ongoing research, development and demonstration projects," *IEEE Power Energ. Mag.*, vol. 5, no. 4, pp. 78–94, 2007.

17 F. Katiraei and M. R. Iravani, "Power management strategies for a microgrid with multiple distributed generation units," *IEEE Trans. Power Syst.*, vol. 21, no. 4, pp. 1821–1831, 2006.

18 C. Hou, X. Hu, and D. Hui, "Hierarchical control techniques applied in micro-grid," in *IEEE Conf. on Power Syst. Tech. (POWERCON)*, Hangzhou, 2010.

19 A. G. Tsikalakis and N. D. Hatziargyriou, "Centralized control for optimizing microgrids" *IEEE Trans. Energy Convers.*, vol. 23, no. 1, pp. 241–248, 2008.

20 A. Khodaei and M. Shahidehpour, "Microgrid-based co-optimization of generation and transmission planning in power systems," *IEEE Trans. Power Syst.*, vol. 28, no. 2, pp. 1582–1590, 2013.

21 R. Lasseter, "Smart distribution: coupled microgrids," *Proc. IEEE*, vol. 99, no. 6, pp. 1074–1082, 2011.

22 A. Ben-Tal and A. Nemirovski. "Robust optimization–methodology and applications," *Math. Program.*, vol. 92, no. 3, pp. 453–480, 2002.

23 H. G. Beyer and B. Sendhoff, "Robust optimization–a comprehensive survey," *Comput. Methods Appl. Mech. Eng.*, vol. 196, no. 33, pp. 3190–3218, 2007.

24 A. Ben-Tal, L. El Ghaoui, and A. Nemirovski, "Robust Optimization," Princeton University Press, 2009.

25 D. Bertsimas, D. B. Brown, and C. Caramanis, "Theory and applications of robust optimization," *SIAM Rev.*, vol. 53, no. 3, pp. 464–501, 2011.

26 T. L. Terry, "Robust linear optimization with recourse: solution methods and other properties," PhD dissertion, University of Michigan, 2009.

27 ILOG CPLEX, "ILOG CPLEX Homepage," [Online]. Available: www.ilog.com, 2009.

6

Engineering Economics of Microgrid Investments

6.1 Principles of Engineering Economics in Microgrids

Microgrids represent a viable alternative to traditional power system expansion by offering shorter construction time and a higher level of system-level reliability [1–4]. The high investment cost of microgrids, however, introduces a major obstacle in the rapid and widespread development of this viable technology. Although microgrids can help to realize lower operations costs in supplying local loads, the microgrid investment cost – which mainly depends on the type, scale, and scope of distributed energy resources (DERs) built in the system using microgrids – is a more decisive factor in microgrid developments. As the microgrid investment cost increases, the microgrid becomes a less attractive option for installation. However, microgrids can create new opportunities for their owners to sell the excess power back to the main grid and be paid at the market price. If the operation cost reduction resulting from the utilization of local DERs and selling excess power to the main grid, along with the benefits from additional improvements and grid services, surpasses the microgrid investment cost, the microgrid planning would be economically justified.

There is extensive research in the literature on short-term microgrid operations planning [5–15]; however, few of them have studied strategic microgrid economics or offered proper models for microgrid operations planning in the long-term horizon. In [16], an operational and planning case study of a hybrid system is carried out to find the best combination of renewable and conventional resources in a microgrid. It identifies the need to account for various combinations of microgrid components to evaluate the optimal operating configurations using advanced planning technologies. In [17], the best location of renewable energy resources within the microgrid is identified using a probabilistic load flow model. In [18], an optimization model for microgrid power generation expansion planning in the context of a low-carbon economy is proposed. An improved matrix real-coded genetic

The Economics of Microgrids, First Edition. Amin Khodaei and Ali Arabnya.
© 2024 The Institute of Electrical and Electronics Engineers, Inc.
Published 2024 by John Wiley & Sons, Inc.

algorithm is employed to solve the planning problem. The results show that clean energy, such as solar and wind, are effective in achieving the targets for CO_2 emission reductions, so it should be advocated and applied more widely, especially in microgrids. In [19], the design, management, and planning of a hydrogen-based microgrid are presented. Experiments are performed on a testbed system. However, the economics of long-term planning was not considered. One of the few models that thoroughly investigates the planning of microgrids along with traditional power system expansion planning is proposed in [20]. In [20], the economic and reliability benefits of microgrids from the system operator's perspective are examined and the interactions of microgrid installations with main grid generation and transmission expansion in a long-term horizon are studied. Furthermore, a microgrid-based co-optimized generation and transmission planning framework is developed to demonstrate the economic and reliability advantages of microgrid investments from the operator's perspective. A comprehensive review on the state of the art in research on microgrids can be found in [23]. Reference [23] highlights the relevant research work on distributed generation, microgrid value propositions, applications of power electronics, economics, operations, control, microgrid clusters, protection, and communication issues. A review of the impact of policy on the economics of microgrid projects can be found in [29].

Microgrids, as enablers of distributed generation, improved grid performance, and the green energy economy, have been deployed on a large scale over the past decade [21] and are expected to continue their growth for the foreseeable future [22]. A significant amount of research has been devoted to studying microgrids and facilitating their development and implementation efforts [23]. The number of published articles in recent years focused on microgrids has increased by multiple folds, and microgrid deployments have been federally supported in the United States, where it has been playing a major role in advancing the nation's energy system [24, 25]. A wide range of applications have been identified for microgrids. The very first modern microgrids were deployed on university campuses, primarily due to the availability of funding for research initiatives and internal expertise in science and engineering [26–28]. Microgrid deployment has become an increasingly attractive solution for commercial and industrial consumers who require premium reliability and power quality levels and are interested in realizing cost savings and economic benefits from the strategic dispatching of their DERs. An example includes prosumers who want to take advantage of available incentives and utilize idle capacity from backup generation to export power to the grid. Moreover, there is a growing number of microgrid deployments in remote locations, as they can be a more efficient, viable alternative for providing electricity services than other solutions, such as upgrading or building new transmission and distribution (T&D) infrastructure. For instance, microgrids can be operated in an interconnected mode to provide peak shaving and defer capacity investments that otherwise would be needed in remote areas, or they can be operated in an islanded mode to supply electricity to isolated areas. The latter approach has been used extensively to provide service to remote rural areas, traditionally via conventional generation (e.g. reciprocating engines and small hydro generation), and more recently via combined dispatch of conventional and renewable DERs. Military microgrids have also received significant traction in recent years, primarily due to support provided by the US Department of Defense. Military microgrids consist of self-sustaining small power grids, which are required to address the mission-critical energy needs of military bases.

More recently, sustainable community microgrids have emerged as an alternative to address the rising societal demands for sustainable electric infrastructure that can provide a high level of reliability and power quality for communities, while being economically viable and environmental-friendly. Sustainable community microgrids aim primarily at supplying electricity from renewable resources for a group of consumers in a neighborhood or several connected neighborhoods in close proximity. Sustainable community microgrids are emerging as a potential solution to address the following trends: (i) residential consumers, who currently use more than one-third of the electric energy produced in the United States [26], are demanding higher levels of reliability and power quality; (ii) utility grids are experiencing an unprecedented proliferation of intermittent renewable DERs, which is motivated by the availability of attractive incentives and regulations designed to address socioeconomic and environmental concerns; (iii) the commercial and industrial consumers' needs to high reliability and quality premium power is growing due to their dependence on sensitive loads; and (iv) the society is demanding more resilient power delivery infrastructure due to the growing dependence on electric service for vital activities such as transportation, e.g. the proliferation of Electric Vehicles (EVs), and the well-documented grid vulnerability issues exposed in recent climate shocks and natural disasters.

The transition from the conventional utility grid to integration with sustainable community microgrids and the enhanced utilization of DERs and controllable loads are anticipated to extensively change the way communities use electricity by increasing energy efficiency, enhancing conservation levels, and lowering greenhouse gas emissions, while lowering the stress level on congested T&D lines. Several obstacles, however, exist in a rapid and widespread deployment of sustainable microgrids, as follows: (i) high capital expenditure (CapEx) requirement for microgrid deployment; (ii) lack of consumer knowledge on potential impacts of DERs and load scheduling strategies; and particularly (iii) ownership and governance structure and regulatory issues.

Despite extensive studies on sustainable microgrids, little literature is dedicated to their engineering economics. This chapter discusses sustainable microgrids and elaborates on different components and anticipated outcomes of these viable deployments. In addition, the economics of microgrid installation from the developer's perspective is investigated. The investment cost of the microgrid is compared with the microgrid economic benefits to justify the microgrid installation. The investment in the microgrid is economically justified when the benefit of utilizing local resources and selling excess power to the main grid surpasses the DER investment cost. The proposed model further compares the candidate DERs to be installed in the microgrid and determines the optimal size and combination of DERs. The proposed model can provide a decision-making framework for electricity customers, including residential and commercial, for economic analysis of new microgrid installations.

6.2 Economic Variables in Microgrid Investments

To perform capital expenditure analysis for microgrid projects, short-term microgrid operations must be integrated with long-term investment. The unit of analysis for investment planning is considered to be annual, while the unit of analysis for the operations cost is considered to be hourly in the planning horizon. For the sake of analysis, it is assumed that the microgrid generation is controllable – i.e. non-dispatchable DERs are coordinated with the energy storage systems to achieve a smooth and dispatchable generation output. In addition, the microgrid installed generation capacity is larger than the microgrid annual consumption peak. This assumption is in line with the microgrid definition that must guarantee the microgrid islanding capability. The microgrid generation is determined based on the market price (i.e. the real-time electricity price in the spot market) received from the system operator. When the local generation price is higher than the market price, the energy is purchased from the system; otherwise, the local generation is utilized. Islanding incidents are considered to be limited; thus, only economic value streams are incorporated into the problem for the sake of simplicity of the illustration of the concept.

The fundamental step in assessing the economic viability of microgrid capital projects is to identify the cost and benefit value streams. The benefit value streams for a sustainable community microgrid project can incorporate a list of variables as follows.

6.2.1 Reliability

One of the most important benefits of sustainable microgrids is to increase power supply reliability. Electric utilities constantly monitor the reliability levels and perform required system upgrades to improve supply availability and to reach and maintain the desired performance. Consumer reliability is typically evaluated in terms of system and customer average interruption frequency and/or duration (e.g. SAIFI, SAIDI, CAIFI, and CAIDI indices, among others). Outages due to storms, equipment failure, and other risk factors impact reliability levels by increasing the average frequency and duration of interruptions. However, when a microgrid is deployed, these metrics can be significantly improved. This is due to the increased redundancy and intrinsic intelligence (control and automation systems) of sustainable community microgrids and the utilization of DERs that allow islanded operations from the grid. Since generation in microgrids is typically located in close proximity to consumer loads, it is less prone to being affected by grid disturbances and infrastructure issues. Additional flexibility to provide service under these conditions is provided by the ability to adjust loads (e.g. demand response) via building and/or microgrid master controllers. Improved reliability can be translated into economic benefits for consumers and utility due to a reduction in interruption costs and energy not supplied (ENS) considering the value of lost load (VOLL). The reliability cost is the amount of ENS multiplied by VOLL during the planning horizon. VOLL is an important measure in power utilities' microeconomics which represents the willingness of customers to pay for their electricity service to avoid curtailment. Valuation of VOLL, which is usually measured in dollars per MWh, can either be based on the marginal value of the next unit of interrupted electricity load or the average value of the interrupted load. However, it varies depending on the type of usage and outage [30]. From an economic point of view, the load interruption cost is considered an opportunity cost. The magnitude of these benefits is dependent upon load criticality and value of the lost load, and also the availability of other alternatives such as backup generation or automatic load transfer trips.

6.2.2 Resiliency

Resiliency refers to the capability of the energy systems to withstand low-probability high-impact events by minimizing possible power outages and quickly returning to the normal operating state with minimum cost. These events typically include extreme weather events and natural and manmade disasters, such as hurricanes, tornados, earthquakes, snow, floods, and cyber and physical attacks, among others. Recent hurricanes in the United States, as well as a documented attack on a California substation [31] and the potentially significant social disruptions, have spawned a great deal of discussion in the power and energy industry about the value and application of microgrids. If the power system is impacted by these events and critical components are severely damaged (e.g. generating facilities and/or T&D infrastructure), service may be disrupted for days or even weeks. The impact of these events on consumers could be mitigated by the deployment of sustainable community microgrids, which allow the local supply of loads even when the supply of power from the utility grid is not available. The increased resilience of the system can be economically quantified through various methods as part of microgrid project benefits and incorporated into the investment analysis.

6.2.3 Carbon Emission Reduction

Microgrids can be a powerful enabler of renewable energy integration to distribution networks. Renewable energy resources may cause significant technical challenges when integrated into a distribution network as they generate a variable supply of energy. Seamless integration of these resources may only be accomplished through the implementation of mitigation measures. Utilities may need to upgrade their distribution network and utilize advanced Volt-VAR control (VVC) practices, smart inverters, or energy storage to address the rapid deployment of renewable energy resources. Community microgrids utilize the coordinated control of a combination of dispatchable DERs, distributed energy storage, and controllable loads to "smooth down" the intermittent output of renewable energy resources. This allows increased penetration of renewable energy and diversification of resources, enables utilities to meet goals set by Renewable Portfolio Standards (RPS), and helps reduce greenhouse gas emissions to combat climate change. Through an internal carbon pricing method, the monetary value of the reduction in carbon emission can be calculated and incorporated into the long-term economic viability assessment of microgrid projects. That includes internal carbon fees, shadow carbon prices, and implicit carbon price methods, among other approaches.

6.2.4 Reduced Costs of Recurring System Upgrades

As the demand for electricity increases, today's power system must be reinforced by the addition of new generation, transmission, and distribution facilities. Sustainable community microgrids deploy DERs to supply local loads, including conventional and renewable DERs, and implement load control to facilitate local grid management. Therefore, while operating in interconnected mode, the additional generating capacity from microgrids can decrease the average and peak system load, effectively deferring capacity increase or generation investments. This benefit, however, is contingent upon a large penetration of microgrids in distribution networks, which could accordingly contribute to mitigating congestion issues. Microgrids can also provide utilities with additional operational flexibility (increased reserve) to handle load transfers during restoration or system reconfiguration. These aspects need to be evaluated when considering microgrid deployment as part of system expansion plans, along with investments in a conventional transmission, distribution, and generation infrastructure.

6.2.5 Energy Efficiency

Microgrids can help improve overall energy efficiency by reducing T&D losses and allowing the implementation of optimal load control and resource dispatch. The former is a direct consequence of supplying consumer loads with local generating facilities, and the latter can be accomplished by intelligent control and dispatch of consumer loads. For instance, the microgrid controller can interact with consumer controllers to curtail or dispatch loads to accomplish overall system efficiency goals. Similarly, this objective could be adjusted to respond to real-time price, operational security, power quality, or reliability signals. Evidently, this requires having the adequate regulatory framework and consumer incentives in place. FERC Order No. 2222 in the United States, to a large extent, has facilitated the regulatory requirements that enable higher energy efficiency by the deployment of DERs through microgrids [32].

6.2.6 Power Quality

Consumers' need for higher power quality has significantly increased during the past decade due to the growing application of voltage-sensitive loads, including a large number and variety of electronic loads and LEDs. Utilities are always seeking efficient ways of improving power quality issues by addressing prevailing concerns stemming from harmonics and voltage. Microgrids provide an efficient answer to address power quality needs by enabling local control of frequency, voltage, load, and the rapid response from distributed energy systems (DES). The financial incentives set by system operators to ensure power quality should be incorporated into the analysis of the economic viability of microgrids.

6.2.7 Lowered Energy Costs

Financial incentives offered to consumers within the microgrid system, who would consider load scheduling strategies according to electricity prices and benefit from locally generated power, are a significant driver in the economic deployment of microgrids. Although it can still be more economical to purchase power from the utilities due to their economies of scale, microgrids can provide benefits by reducing T&D costs. The lowered energy costs impact each individual consumer within the microgrid ecosystem. However, the microgrid local generation not only has the potential to lower energy costs for local consumers (which will be more significant as DER technologies become less expensive), but it could potentially benefit the entire system by reducing the T&D network congestion levels (when the community microgrids penetration is high) and enabling a more economic dispatch of available energy resources in the utility grid.

6.3 Capital Expenditure and Cash Flow Analysis for Microgrids

Microgrid development is economically justified when the expected benefits of their deployment exceed the required CapEx. The microgrid benefits must be comprehensively evaluated and compared with the cost of capital to ensure a complete return on investment to economically justify a microgrid deployment capital project. An accurate assessment of microgrid economic benefits is a challenging task due to the deep uncertainty involved in the data required for its assessment. Moreover, some of the assessment results, such as reliability improvements, are difficult to comprehend for consumers when represented in supply availability terms. Two well-established methods from engineering economics literature for the economic evaluation of capital projects that can be applied to sustainable community microgrid investments include the net present value (NPV) and benefit-cost ratio (BCR) methods. Both methods are related to the concept of net cash flow, i.e. the algebraic sum of expected cash inflows and outflows over a given investment horizon (project's life) for a particular microgrid project. While the NPV method generally provides a more straightforward approach to economically evaluate the projects, the BCR method is generally more desirable for microgrid projects. That is due to the fact that BCR methods would allow the ranking of various investment alternatives, which is often required when there is a constraint on the allocation of resources in the capital budgeting process. The NPV can be calculated as follows [33]:

$$NPV = \sum_{t=0}^{n} \frac{R_t}{(1+i)^t} \tag{6.1}$$

where R_t is the net cash flow in year t, i is the discount rate, and n is the total lifecycle of the project in years. The BCR can be calculated as follows [33]:

$$B/C = \frac{\sum_{j=0}^{n} B_j(1+k)^{-j}}{\sum_{j=0}^{n} C_j(1+k)^{-j}} \tag{6.2}$$

where B_j is the microgrid project benefits in year j, C_j is project costs in year j, and k is the minimum acceptable rate of return (MARR) for the investment. In general, if the project's rate of return is greater than MARR, the project is economically desirable. Thus, efficient planning models are required to ensure the economic viability of microgrid deployments and further justify investments based on cost-benefit analyses under conditions of uncertainty. These benefits, however, must be validated and benchmarked with the microgrid investment cost to ensure the required return on investment to justify their deployment. On the other hand, the project cost stream for a microgrid project should include the required initial CapEx to implement the project, the expected cost of operations and maintenance (O&M) during the lifecycle of the project, and the final disposal cost of the microgrid at the end of its lifecycle. Regardless of the method of choice for engineering economics analysis of a microgrid project, calculation of the net cash flows or project's costs and benefits can be a challenging task considering the interconnection of the microgrid with the main grid, the dynamics of the power market, and the design choices for a given project. Therefore, we propose a CapEx analysis model tailored for microgrids that incorporates the intrinsic nature of microgrid operations into the investment analysis process. First, a microgrid utilizing a single dispatchable DER with a single-step generation price is considered and then the obtained model is extended to a microgrid with multiple DERs.

6.3.1 Microgrid with a Single DER

Assume a microgrid with a single DER and a single-step generation price. The cost of supplying microgrid load is represented by (6.3), as follows:

$$C = \sum_t \lambda P_t + \sum_t \rho_t P_{M,t} \tag{6.3}$$

where ρ_t is the market price at time t and λ is the DER generation price. P_t and $P_{M,t}$ represent the generated power in the microgrid, and power transfer with the main grid at time t, respectively. The total microgrid cost C is the summation of the local generation cost and the cost of purchasing energy from the main grid at the market price overall hours in the scheduling horizon. The local generation is always nonnegative. The main grid power, however, could be positive (import), negative (export), or zero. When the main grid power is negative, i.e. the power is exported from the microgrid to the main grid, the second term would be negative, which represents a benefit for the microgrid rather than a cost. The local generation plus the power purchased from the main grid equals the hourly microgrid load D_t, as shown in (6.4).

$$P_t + P_{M,t} = D_t \tag{6.4}$$

The DER generation, and accordingly the amount of power purchased from the main grid, is determined based on the generation price and market price comparison as represented in (6.5) and (6.6).

$$\lambda \geq \rho_t \Rightarrow P_t = 0, P_{M,t} = D_t \tag{6.5}$$

$$\lambda < \rho_t \Rightarrow P_t = P^{\max}, P_{M,t} = D_t - P^{\max} \tag{6.6}$$

where $P^{\max}$ is the DER maximum capacity. When the market price is lower than the microgrid generation price, the load is supplied from the main grid, and accordingly, the generation of local DER is set to zero. However, when the local generation is less expensive, the DER generates at the maximum power, supplying the microgrid load and selling the excess power to the main grid. Denoting set T_1 as hours in which (6.5) holds true and set T_2 as hours in which (6.6) holds true, the microgrid operation cost (6.3) is rewritten as:

$$C = \sum_{t \in T_1} \rho_t D_t + \sum_{t \in T_2} \lambda P^{\max} + \sum_{t \in T_2} \rho_t \left(D_t - P_m^{\max} \right), \tag{6.7}$$

or in a simpler form:

$$C = \sum_{t} \rho_t D_t + \sum_{t \in T_2} (\lambda - \rho_t) P^{\max} \tag{6.8}$$

The microgrid operation cost in (6.8) includes two terms; the first term is the cost of supplying the microgrid load from the main grid at the entire scheduling horizon, i.e. the cost that customers pay for supplying their load when the microgrid is not installed. The second term is the change in the operation cost of supplying the load when the microgrid is installed. The second term is always negative (since set T_2 includes hours in which $\lambda < \rho_t$); therefore, it characterizes the economic benefit of the microgrid installation.

To justify the microgrid economic development, the benefit should be equal to or larger than the DER investment cost. Considering an annualized investment cost IC for the DER, (6.9) must be satisfied for an economic microgrid development:

$$IC.P^{\max} \leq \sum_{t \in T_2} |\lambda - \rho_t| P^{\max} \tag{6.9}$$

$P^{\max}$ could be removed from both sides of (6.9) to obtain (6.10):

$$IC \leq \sum_{t \in T_2} |\lambda - \rho_t| \tag{6.10}$$

Constraint (6.10) simply states that the positive difference in the market price and microgrid generation price in the planning horizon must exceed the DER investment cost to ensure the microgrid economic development. A graphical representation of this constraint is shown in Figure 6.1, where the market price is shown using an annual price duration curve. Denoting the shaded area by A, the microgrid economic development is justified when $IC \leq A$.

Interestingly, the economic installation of a microgrid does not depend on its size, as represented by (6.10), but on the DER generation price, market price, and the DER investment cost.

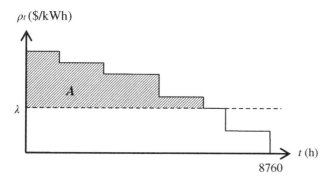

Figure 6.1 Graphical representation of the microgrid economic benefit.

6.3.2 Microgrid with Multiple DERs

DERs have different generation prices and utilize a multistep generation pricing scheme. The uncertainties in load and price forecasts are further considered. The expected cost of supplying microgrid load is obtained in (6.11), which includes the generation cost of DERs inside the microgrid plus the cost of purchasing power from the main grid at the market price. The generation cost of DERs, indexed at i, is summed overall price steps m of DERs at the entire scheduling horizon. The costs are calculated for each scenario s and multiplied by the associated scenario probability p_s. The microgrid load in each scenario and each hour, i.e. D_{st}, is met by the microgrid generation and the energy purchase from the main grid (6.12).

$$C = \sum_s p_s \sum_t \sum_m \sum_i \lambda_{im} P_{imst} + \sum_s p_s \sum_t \rho_{st} P_{M,st} \tag{6.11}$$

$$\sum_m \sum_i P_{imst} + P_{M,st} = D_{st} \qquad \forall s, \forall t \tag{6.12}$$

Assume that the generation steps of DERs are sorted in ascending order. The generation of each DER, as well as the power purchase from the main grid, would be obtained by a comparison between prices (6.13). If a price step is lower than the market price, that step would be dispatched to its maximum; otherwise, the generation of that step would be zero. The purchased power is accordingly obtained from the load balance Eq. (6.12), where a negative value shows that the power is sold back to the utility.

$$\lambda_{im} \geq \rho_{st} \Rightarrow P_{imts} = 0 \tag{6.13}$$

$$\lambda_{im} < \rho_{st} \Rightarrow P_{imts} = P_{im}^{max} \tag{6.14}$$

By replacing (6.13) and (6.14) into (6.11), rearranging the terms, and denoting set T_2 as steps with a price lower than the market price, the total microgrid cost would be as follows:

$$C = \sum_s p_s \sum_t \rho_{st} D_{st} + \sum_s p_s \sum_t \sum_i \sum_{m \in T_2} (\rho_{st} - \lambda_{im}) P_{im}^{\max} \tag{6.15}$$

The first term is the cost of supplying the load when the microgrid is not built (i.e. the entire load is supplied from the main grid). Therefore, the second term is the benefit provided by the microgrid installation (this term is negative and hence regarded as a benefit). Similar to the case with single DER, to ensure microgrid economic development, the microgrid benefit must be larger than the microgrid annualized installation cost over a one-year planning horizon, as follows:

$$\sum_i IC_i P_i^{\max} \leq \sum_s p_s \sum_t \sum_i \sum_{m \in T_2} (\rho_{st} - \lambda_{im}) P_{im}^{\max} \tag{6.16}$$

Each price step could be represented as a percentage of the total DER capacity (6.17), and each DER capacity could be represented as a percentage of the peak load (6.18).

$$P_{im}^{\max} = \gamma_{im} P_i^{\max} \tag{6.17}$$

$$P_i^{\max} = k_i D^{\max} \tag{6.18}$$

where $\sum_m \gamma_{im} = 1$ and $\sum_i k_i = 1$. Accordingly,

$$\sum_i k_i IC_i \leq \sum_s p_s \sum_i \sum_m k_i \gamma_m A_{ims} \tag{6.19}$$

where

$$A_{ims} = \sum_t (\rho_{st} - \lambda_{im}) \qquad \forall (m, i) \in T_2 \tag{6.20}$$

where A_{ims} is the area obtained in Figure 6.1 for each price step m of each individual DER i in scenario s.

The proposed approach is applied to a single-year planning horizon in which annualized investment costs are employed. However, it could be extended to multiyear planning where the total benefit in the entire planning horizon must surpass the total investment cost to ensure the microgrid economic development.

6.4 Case Study

Three candidate DERs are considered for installation in a microgrid. Table 6.1 shows the generation price and annualized investment cost of candidate DERs. The market price is forecasted using historical price data. One hundred scenarios are generated based on the initially forecasted market price using Monte Carlo simulation [34]. The average price duration curve is shown in Table 6.2 where the prices between minimum price, i.e. $0.04/kWh and maximum price, i.e. $0.16/kWh are divided into seven steps with a step size of $0.02/kWh. The microgrid peak annual load is 2 MW.

Using the available information of generation price and market price, the A value (i.e. the shaded area in Figure 6.1) of each DER is calculated as follows: $A_1 = \$32/kW$, $A_2 = \$99.64/kW$, and $A_3 = \$320/kW$. Comparing these values with the investment cost of each DER, the following constraints are obtained:

$$k_1 \times (30 - 32) + k_2 \times (110 - 99.64) + k_3 \times (300 - 320) \leq 0$$
$$k_1 + k_2 + k_3 = 1$$

Table 6.3 represents all possible DER investment combinations based on the obtained constraint. Microgrid installation would be economical with the selection of either DERs 1 or 3. However, individual selection of DER 2 would not justify an economical solution. When selecting DER 2 with any other DER, a constraint, as shown in Table 6.3 have to be satisfied, which sets a limit on the capacity of DER 2. The best option is to install DER 3 as it provides higher economic benefits, compared to its investment cost, for the microgrid.

Table 6.1 DER characteristics.

	DER1	DER2	DER3
Generation price ($/kWh)	0.09	0.06	0.02
Investment cost ($/kW)	30	110	300

Table 6.2 Price duration curve.

ρ_t ($/kWh)	0.16	0.14	0.12	0.10	0.08	0.06	0.04
Duration (h)	21	143	212	908	1456	2769	3251

Table 6.3 DER selection.

Selected DERs	Economical	Constraint
1	Yes	—
2	No	—
3	Yes	—
1 & 2	Yes—conditional	$k_1 \geq 0.84, k_2 \leq 0.16$
1 & 3	Yes	—
2 & 3	Yes—conditional	$k_2 \leq 0.65, k_3 \geq 0.35$
1 & 2 & 3	Yes—conditional	$-2\,k_1 + 10.36\,k_2 - 20\,k_3 \leq 0$

As can be seen, the obtained results do not depend on the size of the microgrid but on the generation price and investment cost of DERs.

Microgrids can provide significant benefits to electricity customers. However, the economic benefits of a microgrid should be compared with the high investment cost to justify its installation. Specific features of the proposed microgrid planning model are listed as follows:

- *DER optimal size and combination:* The proposed models provide a fast and efficient method for determining the optimal size and combination of DERs to be installed in a microgrid.
- *Inclusion of uncertainties:* A stochastic approach is applied to the proposed planning model to incorporate the uncertainty in market price. The number of generated scenarios for the stochastic model represents a trade-off between solution accuracy and execution time. Stochastic planning employs the Monte Carlo simulation for generating scenarios.
- *Simplicity:* The proposed model alleviates the need for complex calculations to ensure the economic development of microgrids.

6.5 Summary

An efficient approach for the economic assessment of microgrid planning is proposed in this chapter. The microgrid cost saving, obtained by the deployment of local resources and selling the excess power to the main grid, is compared with the investment cost to exhibit the economic benefits. Using this model, the optimal mix of DERs to be installed in a microgrid is obtained. The proposed model considered a microgrid with multiple DERs and also multistep price functions.

The computational complexity in the long-term microgrid planning problem is alleviated by a few assumptions, which simplify the calculations. The proposed model is graphically represented, which offers a quick and efficient assessment of the microgrid installation's economic benefits. The model is proposed for microgrids employing a single DER and extended to microgrids with multiple DERs. A stochastic model is used to ensure the practicality of the proposed model. Numerical simulations exhibit the effectiveness of the proposed model. The proposed models can be used for analyzing the microgrid economic development and accordingly, determine the optimal size and combination of DERs to be installed. The proposed model alleviated the need for complex calculations, hence, allowing a fast and efficient study of microgrid planning. The obtained result showed that the economic benefit of a microgrid was independent of the microgrid size. That means, microgrid economics only depends on the characteristics of installed DERs (i.e. annualized investment cost and generation price) and the market price.

References

1 M. Shahidehpour, "Role of smart microgrid in a perfect power system," in *IEEE Power and Energy Soc. Gen. Meeting*, Minneapolis, MN, 2010.

2 F. Katiraei and M. R. Iravani, "Power management strategies for a microgrid with multiple distributed generation units," *IEEE Trans. Power Syst.*, vol. 21, no. 4, pp. 1821–1831, Nov. 2006.

3 M. Shahidehpour, "Global broadcast—transmission planning in restructured systems," *IEEE Power Energ. Mag.*, vol. 5, no. 5, pp. 18–20, Sept./Oct. 2007.

4 S. de la Torre, A. J. Conejo, and J. Contreras, "Transmission expansion planning in electricity markets," *IEEE Trans. Power Syst.*, vol. 23, no. 1, pp. 238–248, Feb. 2008.

5 C. Marnay, F. Rubio, and A. Siddiqui, "Shape of the microgrid," in *IEEE Power Eng. Soc. Winter Meeting*, Columbus, OH, vol. 1, pp. 150–153, 2001.

6 A. Meliopoulos, "Challenges in simulation and design of μgrids," in *IEEE Power Eng. Soc. Winter Meeting*, New York, NY, vol. 1, pp. 309–314, 2002.

7 R. Lasseter, "Microgrids," in *IEEE Power Eng. Soc. Winter Meeting*, New York, NY, vol. 1, pp. 305–308, 2002.

8 Y. Li, D. Vilathgamuwa, and P. Loh, "Design, analysis, and real-time testing of a controller for multibus microgrid system," *IEEE Trans. Power Electron.*, vol. 19, no. 5, pp. 1195–1204, 2004.

9 A. Khodaei, "Microgrid optimal scheduling with multi-period islanding constraints," *IEEE Trans. Power Syst.*, vol. 29, pp. 1383–1392, 2013.

10 C. Hernandez-Aramburo, T. Green, and N. Mugniot, "Fuel consumption minimization of a microgrid," *IEEE Trans. Ind. Appl.*, vol. 41, no. 3, pp. 673–681, 2005.

11 A. Dimeas and N. Hatziargyriou, "Operation of a multiagent system for microgrid control," *IEEE Trans. Power Syst.*, vol. 20, no. 3, pp. 1447–1455, 2005.

12 S. Patra, J. Mitra, and S. Ranade, "Microgrid architecture: a reliability constrained approach," in *IEEE Power Eng. Soc. General Meeting*, San Francisco, CA, pp. 2372–2377, 2005.

13 E. Barklund, N. Pogaku, M. Prodanovic, C. Hernandez-Aramburo, and T. Green, "Energy management in autonomous microgrid using stability constrained droop control of inverters," *IEEE Trans. Power Electron.*, vol. 23, no. 5, pp. 2346–2352, 2008.

14 I. Maity and S. Rao, "Simulation and pricing mechanism analysis of a solar powered electrical microgrid," *IEEE Syst. J.*, vol. 4, no. 3, pp. 275–284, 2010.

15 I. Bae and J. Kim, "Reliability evaluation of customers in a microgrid," *IEEE Trans. Power Syst.*, vol. 23, no. 3, pp. 1416–1422, 2008.

16 W. Su, Z. Yuan, and M. Y. Chow, "Microgrid planning and operation: solar energy and wind energy," in *IEEE Power Energy Soc. Gen. Meeting*, Providence, RI, 2010.

17 O. A. Oke and D. W. P. Thomas, "Probabilistic load flow in microgrid assessment and planning studies," in *IEEE Elect. Power Energy Conf. (EPEC)*, London, ON, Canada, 2012.

18 X. Yang and W. Tian, "Microgrid's generation expansion planning considering lower carbon economy," in *Power Energy Eng. Conf. (APPEEC)*, Shanghai, China, 2012.

19 L. Valverde, F. Rosa, and C. Bordons, "Design, planning and management of a hydrogen-based microgrid," *IEEE Trans. Ind. Inf.*, vol. 9, no. 3, pp. 1398–1404, 2013.

20 A. Khodaei and M. Shahidehpour, "Microgrid-based co-optimization of generation and transmission planning in power systems," *IEEE Trans. Power Syst.*, vol. 28, no. 2, pp. 1582–1590, 2013.

21 M. Shahidehpour and J. Clair, "A functional microgrid for enhancing reliability, sustainability, and energy efficiency," *Electr. J.*, vol. 25, pp. 21–28, Oct. 2012.

22 S. Pandey, J. Han, N. Gurung, H. Chen, E. A. Paaso, Z. Li, and A. Khodaei, "Multi-criteria decision-making and robust optimization methodology for generator sizing of a microgrid," *IEEE Access*, vol. 9, pp. 142264–142275, 2021.

23 S. Parhizi, H. Lotfi, A. Khodaei, and S. Bahramirad, "State of the art in research on microgrids: a review," *IEEE Access*, vol. 3, pp. 890–925, 2015.

24 "The role of microgrids in helping to advance the nation's energy system," May 2022. [Online]. Available: https://www.energy.gov/oe/activities/technology-development/grid-modernization-and-smart-grid/role-microgrids-helping.

25 "The microgrid installation database," May 2021. [Online]. https://doe.icfwebservices.com/microgrid.

26 Business Wire, "$42.3 Bn microgrid markets by connectivity, offering, end use, pattern, type—global forecast to 2026, May 2021. [Online]. Available: www.researchandmarkets.com.

27 "IIT Microgrid," [Online]. Available: http://www.iitmicrogrid.net/microgrid.aspx.

28 E. Krapels, "Microgrid development: good for society and utilities," *IEEE Power Energ. Mag.*, vol. 11, no. 4, pp. 96–94, 2013.

29 K. Milis, H. Peremans, and S. Van Passel, "The impact of policy on microgrid economics: a review," *Renew. Sust. Energ. Rev.*, vol. 81, pp. 3111–3119, 2018.

30 London Economics International LLC, "Estimating the Value of Lost Load," Boston, MA: London Economics International LLC, 2013.

31 R. Smith, US risks national blackout from small-scale attack. *Wall Street J.*, vol. 12, 2014.

32 E. D. Cartwright, "FERC order 2222 gives boost to DERs," *Clim. Energy*, vol. 37, no. 5, pp. 22–22, 2020.

33 G. T. Stevens, "The Economic Analysis of Capital Expenditures for Managers and Engineers," Ginn Press, 1992.

34 H. Y. Yamin, M. Shahidehpour, and Z. Li, "Adaptive short-term electricity price forecasting using artificial neural networks in the restructured power markets," *Int. J. Electr. Power Energy Syst.*, vol. 26, no. 8, pp. 571–581, 2004.

7

Microgrid Planning Under Uncertainty

7.1 Dynamics of Uncertainty in Microgrids

The need for effectively managing the growing penetration of distributed energy resources (DERs) in electric power systems as well as the increasing expectations for enhancing the power system's resilience, reliability, and quality in the wake of climate change and cyber-physical threats necessitate a fundamental change in system design by decentralizing the grid through adoption of microgrids. Representing small-scale power systems, microgrids would generate, distribute, and regulate the flow of electricity to local customers with a high degree of flexibility and efficiency on both supply and demand sides [1–5]. Despite the technological advantages that it will bring to the power grid, the relatively large capital expenditure required for the deployment of microgrids can serve as a major obstacle for the large-scale adoption of this viable technology. Therefore, economic solutions and financial innovations can play a major role in unleashing the potential of this game-changing technology in the era of climate change and cyber-physical security threats, given their increasing risk landscape.

Microgrids can provide significant advantages to local customers and to the main grid. In a short-term planning horizon, microgrids can improve power quality by managing local loads and regulating voltage, reduce carbon emission by diversifying energy mix with an increasing share of renewable resources, reduce generation costs by the employment of less expensive renewable energy resources such as solar and wind, reduce transmission and distribution costs by offering a local generation rather than purchasing power from the main grid, and enable a significant level of energy efficiency by dynamically responding to real-time market prices [6–9]. In the long-term horizon, microgrids can improve the system's reliability by enabling self-healing mechanisms at local distribution networks, offer viable alternatives to conventional power system expansion models by eliminating concentrated, large-scale capital project investments in generation and

The Economics of Microgrids, First Edition. Amin Khodaei and Ali Arabnya.
© 2024 The Institute of Electrical and Electronics Engineers, Inc.
Published 2024 by John Wiley & Sons, Inc.

transmission infrastructure, and provide a wide range of long-term economic benefits when they are co-optimized with generation and transmission expansion planning [10, 11]. These advantages can help the microgrid technology to stand out as a promising enabler of transitioning to a decentralized, decarbonized, and digitalized power grid of the future and can unleash a wide range of economic and financial solutions for mobilization and allocation of capital that are required for a power and energy infrastructure of the twenty-first century.

A substantial amount of investment in microgrids has been made during the past few years. The global microgrid power capacity is estimated at 24,217 GW as of the first quarter of 2022 [12]. Extensive research work has been conducted on the short-term operations and control of microgrids [8, 13–22]. The communication aspects of microgrids are also broadly studied [23, 24]. However, the study on the long-term planning of microgrids is very limited. In [25] an operations and planning case study of a hybrid system is carried out to find the best combination of renewable and conventional resources in microgrids. It concludes that various combinations of microgrid components must be considered to evaluate the optimal operating configurations. In [26], the best locations of renewable energy resources within the microgrid are identified using a probabilistic load flow model. In [27], an optimization model for microgrid power generation expansion planning considering a low-carbon economy is proposed. An improved matrix real-coded genetic algorithm is employed to solve the planning problem. The results show that clean energy resources, such as solar and wind, are effective in achieving CO_2 emission reduction targets. In [28], the design, management, and planning of a hydrogen-based microgrid are presented. Experiments are performed on a testbed; however, the long-term planning economics are not taken into account. In [11], the microgrid planning is co-optimized with the generation and transmission expansion planning. The economic and reliability benefits of the microgrid from a system operator's perspective are examined, the interaction of microgrid installations with main grid generation and transmission expansion is investigated, and the optimal locations of microgrid installations in the power network are determined.

The microgrid planning problem is significantly different from the generation expansion problem in conventional power systems. In the generation expansion problem, the optimal combination of generation resources to be added to the system for satisfying the forecasted load growth is determined. Thus, the new facilities are installed to complement the existing infrastructure and improve economic and reliability performance. In microgrid planning, however, the installation decision is made by developers as a standalone project regardless of its impact on the main grid. In contrast to conventional power systems, microgrids include a high percentage of variable renewable generation resources, energy storage systems, and adjustable loads. The uncertainty of these technologies adds an additional layer of complexity to the operations planning in microgrids. Moreover, in the generation expansion planning of conventional power systems, a sufficient level of generation reserve margin must be considered to account for contingencies and unexpected outages. In microgrids, however, they are typically operating in the grid-connected mode, in which the main grid acts as a de facto infinite bus that provides the required reserve. Microgrids can occasionally be disconnected from the main grid, primarily due to disturbances in the distribution network, and can operate in the islanded mode. Moreover, in power systems, the timing of installation is an important factor, in which a delayed investment in a candidate resource provides economic benefits to the system. In microgrids, all the components are usually installed during the first year of the planning horizon, where additional investments are limited to working capital to cover operations and maintenance (O&M) costs. Both problems, however, share a common objective, which is to minimize the total lifecycle cost, i.e. the cost of initial capital expenditure plus the net present value (NPV) of the operational cost. These differences preclude a direct application of conventional power system planning models to the microgrid planning problem, thus new methods are required to address this problem.

In this chapter, the problem of microgrid planning under uncertainty is investigated from an owner's perspective. We introduce a novel model to determine the optimal size and combination of DERs to be installed in the microgrid to guarantee that the investment cost will be recovered. DERs in the microgrid require a relatively high capital expenditure, whereas the generation cost of these resources can be lower than the spot price in electricity markets. When the generation price of a DER in the microgrid is less than the market price, it will create an incentive to sell excess power to the main grid and be renumerated at a market price. If the total expected profit is estimated to be higher than the capital expenditure, it would economically justify the microgrid installations. Prevailing planning and operations constraints are considered for ensuring microgrid islanding ability and accounting for future load growth. The solution is improved by considering the real-world uncertainties associated with the planning problem using a stochastic approach that considers multiple probabilistic scenarios. Moreover, we show how the load and price prediction errors can be reduced using duration curves. The proposed model can provide an efficient tool for developers and electricity customers to assess the economic feasibility of their microgrid installations.

7.2 Economics of Uncertainty in Microgrids

DERs installed in a microgrid are typically subject to a high initial capital expenditure and a low operational cost [10, 29–31]. The typical microgrid DERs require a higher capital investment compared to conventional energy resources. The DERs in the microgrid, however, can provide a less expensive generation compared to the energy purchased from the main grid, particularly at times of transmission network congestion. The benefit from selling energy to the main grid is considered the major source of payback to recover microgrid capital expenditure.

Microgrid planning requires prediction of the loads and the market price for the entire planning horizon. Predictions, however, can involve a high degree of error as several uncertain factors are involved in the prediction process. Particularly, the market price, in addition to bids made by generation companies for power generation, the transmission network congestion and losses also play important roles in price discovery in the market. Predicting system behavior can be more challenging when adding microgrids and responsive loads with the ability to respond to market price variations, as load adjustments can significantly impact market prices [32]. To reduce the prediction errors and effectively model the market price variations, we employ price duration curve (PDC) and load duration curve (LDC). The PDC measures the average probability that the hourly price will exceed a certain quantity x, as follows:

$$PDC_s(x) = \frac{1}{T}\sum_t prob(\rho_{st} > x) \tag{7.1}$$

$$LDC_s(x) = \frac{1}{T}\sum_t prob(D_{st} > x) \tag{7.2}$$

where PDC_s is the PDC in scenario s, ρ_{st} is the real-time market price in scenario s at time t, LDC_s is the LDC, and, D_{st} is the real-time load in scenario s at time t. The PDC ranks market prices during the planning horizon in descending order, hence reducing the error in the prediction of hourly market prices. The PDC and LDC are generated for each scenario to further capture the uncertainty in the price and load predictions. It is assumed that the load and market price in each hour follow a normal distribution with a quantifiable mean and standard deviation [33]. The PDC is employed in the microgrid planning problem in an hourly fashion (i.e. 8760 hours per year). The PDC can be represented using price blocks. The choice of blocks instead of hourly values, however, should consider the tradeoff between the solution accuracy and the computational efficiency of the proposed model.

7.3 An Economic Model for Microgrid Planning Under Uncertainty

Consider a set of candidate DERs in a microgrid system. The microgrid is operated in the grid-connected mode, which can switch to the islanded mode to satisfy local loads in case of main grid disturbances. It is assumed that all candidate DERs are installed during the first year of the planning horizon. Without loss of generality, the microgrid network congestion is neglected, which eliminates the need for determining the optimal location of DERs. Therefore, the microgrid planning problem determines the optimal size and combination of candidate DERs.

7.3.1 Microgrid Planning Objective

The objective of the microgrid planning problem consists of the required DER investment plus the expected cost of microgrid operations, as follows:

$$\text{Min } C = \sum_i CC_i P_i^{\max} + \sum_s p_s \sum_t \sum_i \sum_q \lambda_{qi} P_{m,qist} + \sum_s p_s \sum_t \rho_{st} P_{g,st}, \quad (7.3)$$

where i is the index for DERs, s is the index for scenarios, t is the index for time, C is the microgrid total planning cost, CC_i is the annualized investment cost of DER i, p_s is the probability of scenario s, $P_i^{\max}$ is the maximum capacity of microgrid i, $P_{g,st}$ is main grid power transfer in scenario s at time t, $P_{m,qist}$ is the generation of step q of DER i in scenario s at time t, $P_b^{\max}$ is the maximum capacity of the energy storage system, λ_{qi} is the price of step q of DER i, and ρ_{st} is the real-time market price in scenario s at time t. The selected DERs would be installed during the first year of the planning horizon. The microgrid's expected operations cost includes the generation cost of DERs in the microgrid plus the cost of purchasing power from the main grid. The DER dispatch and generation costs are obtained from a linearized, multi-step generation–price curve. The cost of power purchase is defined as the market price at the point of common coupling multiplied by the amount of purchased power. The expected costs are calculated by considering multiple operation scenarios and summing over probability-weighted costs in each scenario. Note that main grid power in (7.3), i.e. $P_{g,st}$, can be negative, which indicates the microgrid is selling excess power to the main grid and is paid at the market price.

7.3.2 Planning Constraints

The installed generation capacity must be larger than the annual peak load so the microgrid would supply the local load without interruption when disconnected from the main grid distribution network. The peak load is obtained over all scenarios in the planning year, i.e. $P_D^{\max} = \max_{s,t} \{P_{D,st}\}$. This requirement can be modeled through following constraints:

$$\sum_{i \notin I_n} P_i^{\max} \geq (1 + R) P_D^{\max} \qquad (7.4)$$

$$\sum_i CC_i P_i^{\max} \leq Inv^{\max} \qquad (7.5)$$

where $P_i^{\max}$ is the maximum capacity of microgrid i, R is microgrid load growth percentage, $P_D^{\max}$ is maximum microgrid load, I_n is a set of non-dispatchable units, and $Inv^{\max}$ is the annual investment budget. It is likely that the non-dispatchable generation is not sufficient to support disconnection from the main grid at the time of islanding, so the non-dispatchable units are excluded from the microgrid's installed generation capacity, as modeled in (7.4). The load is multiplied by a constant R to reflect future load growth. The microgrid investment cost is limited by the available annual budget, as modeled in (7.5).

7.3.3 Operational Constraints

The planning objective is subject to a set of operational constraints, as follows:

$$\sum_i P_{m,ist} + P_{b,st} + P_{g,st} = P_{D,st} \qquad \forall s, \forall t \tag{7.6}$$

$$P_{m,ist} \in \Omega_i \qquad \forall i, \forall s, \forall t \tag{7.7}$$

$$P_{b,st} \in \Omega_b \qquad \forall s, \forall t \tag{7.8}$$

$$\sum_t P_{b,st} = 0 \qquad \forall s \tag{7.9}$$

$$\sum_i P_{it} + P_{M,t} = \sum_d D_{dt} \qquad \forall t \tag{7.10}$$

$$-P_M^{\max} \leq P_{M,t} \leq P_M^{\max} \qquad \forall t \tag{7.11}$$

$$P_i^{\min} I_{it} \leq P_{it} \leq P_i^{\max} I_{it} \qquad \forall i \in G, \forall t \tag{7.12}$$

$$P_{it} \leq P_{it}^{dch,\,\max} u_{it} - P_{it}^{ch,\,\min} v_{it} \qquad \forall i \in S, \forall t \tag{7.13}$$

$$P_{it} \geq P_{it}^{dch,\,\min} u_{it} - P_{it}^{ch,\,\max} v_{it} \qquad \forall i \in S, \forall t \tag{7.14}$$

$$u_{it} + v_{it} \leq 1 \qquad \forall i \in S, \forall t \tag{7.15}$$

$$D_{dt}^{\min} z_{dt} \leq D_{dt} \leq D_{dt}^{\max} z_{dt} \qquad \forall d \in D, \forall t \tag{7.16}$$

$$\sum_{t \in [\alpha_d, \beta_d]} D_{dt} = E_d \qquad \forall d \in D \tag{7.17}$$

where the load balance is represented by (7.6), where $P_{m,ist}$ is the generation of DER i in scenario s at time t, $P_{b,st}$ is the generation/consumption of energy storage system in scenario s at time t, $P_{g,st}$ is the main grid power transfer in scenario s at time t, and $P_{D,st}$ is microgrid load in scenario s at time t, where the sum of local generation, energy storage generation/consumption, and main grid power transfer matches the microgrid load at any given time in each scenario. The local generation and load are always nonnegative; however, the energy storage power and main grid power could be negative, positive, or zero, depending on the direction of power flow. When there is a contingency in the main grid, the microgrid would be transferred from grid-connected to islanded mode to ensure supply reliability, so the main grid power would be zero. The local generation comes from dispatchable and non-dispatchable units, which are subject to prevailing operational constraints depending on the type of DER (7.7) where Ω_i is a set of operational constraints for DER i, which include but is not limited to generation capacity limits, ramping up/down limits, minimum up/downtime limits, and fuel and emission limits. The storage limits are considered in (7.8), which include charging/discharging limits and state of charge limits, where Ω_b is a set of operational constraints for energy storage system.

The application of duration curves would significantly simplify the modeling of energy storage systems. The energy storage charging and discharging modes are time-dependent and are governed by the amount of power available to be discharged or the available capacity to further charge the energy storage. Using duration curves, the charging and discharging of the energy storage is not chronological; therefore, alongside the energy storage limits in (7.8) should be satisfied. Equation (7.9) ensures a practical energy storage operation and guarantees equal charged and discharged energy in the planning year. The generation of dispatchable units, however, is obtained based on economic considerations. Figure 7.1 depicts a multi-step generation cost curve for a dispatchable unit. To determine the unit generation, the price of each step is compared with the market price. If the marginal cost of a step is smaller than the market price, that step would be

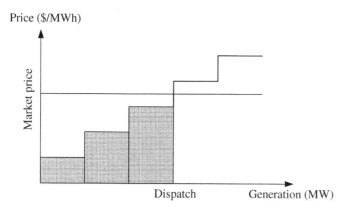

Figure 7.1 Calculating dispatch of dispatchable units.

dispatched to its maximum; otherwise, the generation of that step would be zero. Note that when the market price is higher than the marginal cost of all the steps, the DER would be dispatched at its maximum capacity. Given the market price, Figure 7.1 shows the dispatchable unit dispatch based on generation steps. A higher number of steps would result in a better approximation of the quadratic generation cost function, while imposing an insignificant computational burden on the problem.

The proposed microgrid planning model can be decomposed into several smaller and easier-to-solve subproblems, representing a master planning problem and several operation subproblems for each scenario. However, the proposed model has a small size, where only a few DERs are considered for installation. Furthermore, the microgrid planning problem is a long-term problem, where computational speed is not a barrier.

7.3.4 Economic Assessment of DER Selection

The objective function in (7.3) can be rewritten by substituting the value of $P_{g,st}$ from (7.6). By rearranging the terms, the objective function can be rewritten as follows:

$$
C = \sum_s p_s \sum_t \rho_{st} P_{D,st}
$$
$$
+ \left\{ \sum_i CC_i P_i^{\max} + \sum_s p_s \sum_t \sum_i \sum_q (\lambda_{qi} - \rho_{st}) P_{m,qist} - \sum_s p_s \sum_t \rho_{st} P_{b,st} \right\},
$$
$$
(7.18)
$$

where the first term is the expected annual cost of supplying microgrid load when the required power is fully supplied from the main grid, i.e. the total cost that loads would pay if the microgrid was not installed. The terms in curly brackets represent the total cost associated with the microgrid installation, which includes the annual investment cost and the cost of DERs. When the microgrid is installed, the second term in (7.18) is added to the objective, which quantifies the cost/benefit provided by the microgrid. If this term is negative, it indicates that microgrid installation would reduce the expected cost of supplying the local loads. Any set of candidate DERs that ensures a negative value, i.e. a benefit, is considered to be an economically justified solution to the microgrid planning problem. On the other hand, if this term is positive and the microgrid installation adds additional cost to the cost of supplying the load, it would not be considered an economically justified solution. The solution obtained in the previous subsection is the solution with the minimum possible cost, which provides the highest benefit.

A microgrid requires investment in distribution network improvement, smart switches, sensors, and measurement devices, as well as master and local controllers, among others. However, as the largest portion of the microgrid investment must be allocated to installing DERs, this model, without loss of generality, only focuses on costs related to DERs. Note that the other costs can be added as a constant to the objective function of the microgrid planning problem.

7.4 Case Study

Consider a microgrid for installation with a group of electricity customers with a peak annual load demand of 10 MW. The future load growth is assumed to be 20% of the current load. Five candidate DERs are considered for installation, including four distributed generation units and one energy storage system, as represented in Table 7.1. The DERs present a tradeoff between the investment cost and generation cost, where the DER with a higher investment cost has a lower generation cost and vice versa. In Table 7.1, DER 1 and DER 2 are dispatchable, while DER 3 and DER 4 are non-dispatchable units, and DER 5 is an energy storage system.

7.4.1 PDC vs. Chronological Curve

Figure 7.2 depicts the real-time market price and the predicted price for the duration of one day represented by PDC and chronological curve. The predicted, real-time prices are obtained from operations data from a real-world power system [34]. In a one-day prediction, the PDC method results in a mean absolute error (MAE) of 15.4%, compared to an MEA of 18.7% for the chronological curve. Therefore, using the PDC model significantly reduces the prediction error. This reduction will be even more significant when longer timespans are considered. Figure 7.3 depicts the chronological curve and PDC of real-time and predicted prices for the duration of one week. Using PDC, the prediction MAE is reduced from 55.4% to 31.0%. Therefore, the obtained curve offers more than 24% prediction improvement over

Table 7.1 DER characteristics.

DER number	Maximum size (MW)	Annualized investment cost ($/MW)	Generation cost ($/MWh)
1	8	20,000	70
2	8	70,000	40
3	8	120,000	20
4	8	250,000	0
5	5	60,000	0

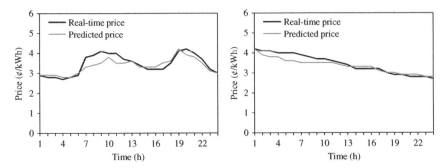

Figure 7.2 Comparison of predicted vs. real-time prices for a one-day period (left: chronological curve; right: PDC).

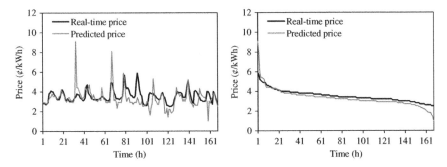

Figure 7.3 Comparison of predicted vs. real-time prices for a one-week period (left: chronological curve; right: PDC).

the chronological curve. Figure 7.4 compares the real-time and predicted prices for the chronological curve and PDC for a duration of one month. The MAE of PDC gains an improvement of more than 32% from the chronological curve with an error of 60%. This trend can be extended to longer time periods, where the MAE for the PDC method is significantly less than the chronological curve method. Figures 7.3–7.5 illustrate that the PDC model offers a significantly lower prediction error compared to the chronological model, thus providing a more trustworthy price prediction for microgrid planning.

Unlike chronological price curve, PDC can be represented by price blocks. As defined in (7.1), the PDC measures the average probability that the hourly price will exceed a certain quantity x. By changing the intervals in x values, price blocks are obtained. The selection of price blocks represents a tradeoff between solution accuracy and computational speed. Further, the MAE of PDC with blocks is

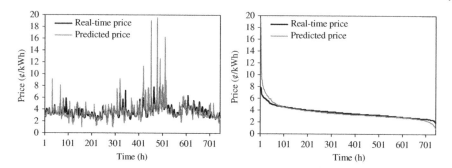

Figure 7.4 Comparison of predicted vs. real-time prices for a one-month period (left: chronological curve; right: PDC).

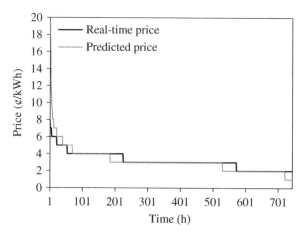

Figure 7.5 Block representation of predicted price and real-time price for a one-month period.

smaller than the original PDC. Figure 7.5 shows the real-time and predicted prices represented by price blocks. An interval of 1¢/kWh between x values is assumed.

The mean and standard deviation of the price for the considered load are calculated from historical data at the point of common coupling as 3.6407, and 1.5401, respectively. A normal distribution is fitted, and random numbers are sampled from this distribution to represent the market price. A total of 100 scenarios are generated for each hour of a one-year planning horizon, each with a probability of 0.01 to account for the uncertainty associated with the prices. The obtained price profile is accordingly converted to PDC using (7.2), where x ranges from the minimum to the maximum value of the generated scenario for prices with a step of

0.0001¢/kWh. A similar procedure is performed to obtain LDC in different scenarios based on the historical load data from a real-world microgrid. The mean and the standard deviation of the load are obtained as 6.002, and 1.1527, respectively.

7.4.2 Optimal Microgrid Planning

The proposed microgrid planning problem is solved to determine the optimal size and combination of candidate DERs. We assume no constraints on the capital investment budget for the microgrid project. The derived optimal solution is to install DER 2 (dispatchable unit), DER 3 (non-dispatchable unit), and DER 5 (energy storage system), which have the capacities of 8 MW, 8 MW, and 5 MW, respectively. The total microgrid planning cost is $2,040,511, including annual investment cost of $1,820,000, DER generation cost of $2,433,406, and the cost of −$2,212,900 for the main grid. The negative cost for the main grid represents the expected economic benefit (profit) that can be obtained by selling excess power from microgrid to the main grid. With a counterfactual analysis, if the microgrid is not installed, the cost of supplying the load would have been $2,478,398 – which shows that the microgrid installation provides an annual saving of $437,887 (=$2,478,398 − $2,040,511). The installation of DERs 2 and DER 3 illustrate that the DER installation is a function of both microgrid investment cost and generation cost. While DER 1 has a very low required capital expenditure and the generation of DER 4 is zero, these units do not provide the same level of economic benefits that can be realized by DERs 2 and DER 3. Moreover, the energy storage system (DER 5), which acts as a load for half of the planning year, provides more economic benefits than DER 1 and DER 4.

The size of DERs and the resulting benefit would be confined when the microgrid investment is restricted by an annual capital expenditure budget. Figure 7.6 shows the total microgrid planning cost as a function of microgrid' annual investment budget. For budgets less than $0.3 M, the microgrid installation is not economically justified; therefore, the total planning cost would be equal to the cost of supplying the load from the main grid. Higher budgets will justify microgrid installation as the benefit obtained from microgrid is increased. A higher budget will result in a lower total planning cost, whereas the drop in total planning cost is significant for budgets up to $0.9 M. On the other hand, higher budgets will increase the total investment cost. On the other hand, Figure 7.6 represents a straight line with a slope of 1 for the total capital expenditure with budget constraints ranging from $0.4 million to $1.82 million. As shown, the entire budget is used up to install the DERs. For budget constraints of higher than $1.82 M, however, the total investment cost does not increase. At this point, the economical DERs have reached their maximum capacity; hence, a higher budget would not result in additional DER installation. Figure 7.7 exhibits that a tradeoff exists between the available budget

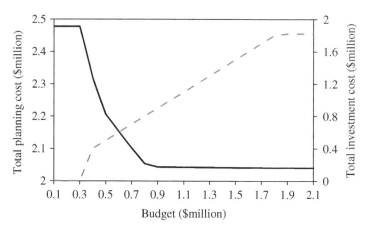

Figure 7.6 Total planning cost (solid line) and total DER investment (dashed line) as a function of investment budget.

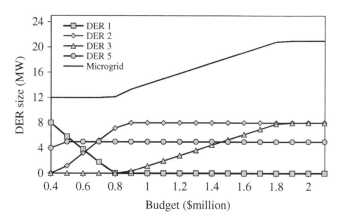

Figure 7.7 DER size as a function of microgrid investment budget.

and the cost of microgrid installation. This sensitivity analysis on budget is critical to prevent underinvestment and overinvestment decisions in microgrid projects. Figure 7.7 shows the DER capacity and combination as a function of budget. As the budget increases, the DER capacity and combination change as well. For low budget limits, DER 1 is more economical as it offers a lower investment cost. However, as the budget ceiling increases, DER 2 and DER 3 become economically more attractive options. DER 4 is not installed and thus is not shown in the figure. A higher budget ceiling will result in additional DER investments and benefits. It is evident that the budget ceiling can directly control the optimal DER capacity and combination and subsequently influence the economic benefit and capital expenditure.

While microgrids can deliver significant benefits to electricity customers in terms of reliability and power quality, the economic benefits of a microgrid system should be assessed to justify the relatively large investment in DERs. The specific features of the proposed microgrid planning model are as follows:

- *Optimal DER selection:* The proposed model provides an accurate and efficient model to determine the optimal capacity and combination of DERs to be installed in a microgrid to maximize its economic benefits. The choice of DERs in a microgrid is determined by DER capital expenditure, generation cost, budget constraints, and the market price for the energy.
- *Incorporating the uncertainties:* A stochastic approach is applied to the proposed planning model to incorporate the uncertainty in price and load predictions. An adequately large number of scenarios can be generated and incorporated into the proposed model to account for uncertainties in a real-world system.
- *Planning horizon considerations:* The short-term operation and long-term planning are considered simultaneously. The short-term problem includes hourly operation of DERs and microgrid-main grid interaction, while the long-term problem incorporates DER investments.
- *Computational efficiency:* The size of candidate DERs is considered as a variable in the proposed model, ranging from zero to the maximum available capacity for a candidate DER using a linear model. Using this linear model, the need for additional binary variables to represent the microgrid installation state is eliminated.
- *Price and load prediction:* A PDC model is employed to significantly reduce the error in long-term price predictions. Similarly, an LDC is employed for load predictions. The duration curves are suitable for the planning model and could further reduce the computational burden when represented in blocks.
- *Annualized model:* The DERs in a microgrid are assumed to be installed in the first year of microgrid deployment, thus facilitating the utilization of an annualized planning model. Relevant constraints are introduced to maintain a higher installed capacity than the peak load, to allow islanding, and to reflect possible load growth in the future.

7.5 Summary

We introduced a stochastic microgrid planning model to determine the optimal capacity and combination of DERs for a microgrid system. The proposed model minimizes the total cost during the planning horizon, including the cost of capital for DER investment as well as the cost of supplying the load. The cost saving, realized by employing local resources and selling the excess power to the main grid, should recover the relatively high amount of capital expenditure for installation of DERs in order to economically justify the deployment of a microgrid project. Price and LDCs are employed instead of chronological curves to reduce prediction errors. The reduced error would guarantee a more accurate solution to the planning problem. The duration curves show great applicability to the planning problem and guarantee a substantial error reduction for longer prediction periods. Numerical case study demonstrates the effectiveness of the proposed model. Further, we analyzed the sensitivity of the microgrid planning to the budget constraint for additional insights needed to avoid misallocation of capital by underinvesting or overinvesting in microgrid projects. The model demonstrates that the DER selection was not only a function of DERs capital and generation costs but also of the budget ceiling on maximum allowable investment in a microgrid project.

References

1 M. Shahidehpour, "Role of smart microgrid in a perfect power system," in *IEEE Power and Energy Society General Meeting*, Minneapolis, MN, 2010.

2 A. Flueck, Z. Li, "Destination perfection," *IEEE Power Energ. Mag.*, vol. 6, no. 6, pp. 36–47, Nov./Dec. 2008.

3 M. Shahidehpour and J. Clair, "A functional microgrid for enhancing reliability, sustainability, and energy efficiency," *Electr. J.*, vol. 25, pp. 21–28, Oct. 2012.

4 S. Bahramirad, W. Reder, and A. Khodaei, "Reliability-constrained optimal sizing of energy storage system in a microgrid," *IEEE Trans. Smart Grid*, vol. 3, no. 4, pp. 2056–2062, Dec. 2012.

5 N. Hatziargyriou, H. Asano, M. R. Iravani, and C. Marnay, "Microgrids: an overview of ongoing research, development and demonstration projects," *IEEE Power Energ. Mag.*, vol. 5, no. 4, July/Aug. 2007.

6 A. G. Tsikalakis, N. D. Hatziargyriou, "Centralized control for optimizing microgrids operation," *IEEE Trans. Energy Convers.*, vol. 23, no. 1, Mar. 2008.

7 B. Kroposki, R. Lasseter, T. Ise, S. Morozumi, S. Papathanassiou, and N. Hatziargyriou, "Making microgrids work," *IEEE Power Energ. Mag.*, vol. 6, no. 3, May 2008.

8 I. Bae and J. Kim, "Reliability evaluation of customers in a microgrid", *IEEE Trans. Power Syst.*, vol. 23, no. 3, 1416–1422, Aug. 2008.

9 S. Kennedy, M. Marden, "Reliability of Islanded Microgrids with Stochastic Generation and Prioritized Load," Bucharest: IEEE Powertech, June 2009.

10 Federal Energy Management Program, "Using Distributed Energy Resources, A How-To Guide for Federal Facility Managers," DOE/GO-102002-1520, US Department of Energy, May 2002.

11 A. Khodaei and M. Shahidehpour, "Microgrid-based co-optimization of generation and transmission planning in power systems," *IEEE Trans. Power Syst.*, vol. 28, no. 2, pp. 1582–1590, May 2013.

12 Microgrid Deployment Tracker 1Q22; Guidehouse Insights, 2022

13 C. Marnay, F. Rubio, and A. Siddiqui, "Shape of the microgrid," in *IEEE Power Engineering Society Winter Meeting*, Columbus, OH, vol. 1, pp. 150–153, 2001.

14 A. Meliopoulos, "Challenges in simulation and design of µgrids," in *IEEE Power Engineering Society Winter Meeting*, vol. 1, pp. 309–314, 2002.

15 R. Lasseter, "Microgrids," in *IEEE Power Engineering Society Winter Meeting*, New York, NY, vol. 1, pp. 305–308, 2002.

16 Y. Li, D. Vilathgamuwa, and P. Loh, "Design, analysis, and real-time testing of a controller for multibus microgrid system," *IEEE Trans. Power Electron.*, vol. 19, no. 5, pp. 1195–1204, 2004.

17 Y. Zoka, H. Sasaki, N. Yorino, K. Kawahara, and C. Liu, "An interaction problem of distributed generators installed in a microgrid," in *IEEE International Conference on Electric Utility Deregulation, Restructuring and Power Technologies*, Hong Kong, China, vol. 2, pp. 795–799. IEEE, 2004.

18 C. Hernandez-Aramburo, T. Green, and N. Mugniot, "Fuel consumption minimization of a microgrid," *IEEE Trans. Ind. Appl.*, vol. 41, no. 3, pp. 673–681, 2005.

19 A. Dimeas and N. Hatziargyriou, "Operation of a multiagent system for microgrid control," *IEEE Trans. Power Syst.*, vol. 20, no. 3, pp. 1447–1455, 2005.

20 S. Patra, J. Mitra, and S. Ranade, "Microgrid architecture: a reliability constrained approach," in *IEEE Power Engineering Society General Meeting*, San Francisco, CA, pp. 2372–2377, 2005.

21 E. Barklund, N. Pogaku, M. Prodanovic, C. Hernandez-Aramburo, and T. Green, "Energy management in autonomous microgrid using stability constrained droop control of inverters," *IEEE Trans. Power Electron.*, vol. 23, no. 5, pp. 2346–2352, 2008.

22 I. Maity and S. Rao, "Simulation and pricing mechanism analysis of a solar powered electrical microgrid," *IEEE Syst. J.*, vol. 4, no. 3, pp. 275–284, 2010.

23 S. Galli, A. Scaglione, and Z. Wang, "Power line communications and the smart grid," in *Proc. International Conference on Smart Grid Communications*, Gaithersburg, MD, Oct. 2010.

24 G. Bag, R. Majumder, and K. Kim, "Low cost wireless sensor network in distributed generation," *International Conference on Smart Grid Communications*, Gaithersburg, MD, Oct. 2010.

25 W. Su, Z. Yuan, and M.-Y. Chow, "Microgrid planning and operation: solar energy and wind energy," *Power and Energy Society General Meeting*, Minneapolis, MN, pp. 1–7, IEEE, 25–29 July 2010.

26 O. A. Oke and D. W. P. Thomas, "Probabilistic load flow in microgrid assessment and planning studies," in *Electrical Power and Energy Conference (EPEC)*, London, ON, Oct. 2012.

27 X. Yang and W. Tian, "Microgrid's generation expansion planning considering lower carbon economy," in *Power and Energy Engineering Conference (APPEEC)*, Shanghai, China, Mar. 2012

28 Valverde, L.; Rosa, F.; Bordons, C., "Design, planning and management of a hydrogen-based microgrid," *IEEE Trans. Industr. Inform.*, vol. 9, pp. 1398–1404, 2013.

29 F. Katiraei, M. R. Iravani, "Power management strategies for a microgrid with multiple distributed generation units," *IEEE Trans. Power Syst.*, vol. 21, no. 4 Nov. 2006.

30 C. Hou X. Hu, and D. Hui. "Hierarchical control techniques applied in micro-grid," in *IEEE Conf. on Power Syst. Tech. (POWERCO)*, Hangzhou, Oct. 2010.

31 A. G. Tsikalakis and N. D. Hatziargyriou, "Centralized control for optimizing microgrids" *IEEE Trans. Energy Convers.*, vol. 23, no. 1, pp. 241–248, 2008.

32 A. Khodaei, M. Shahidehpour, and S. Bahramirad, "SCUC with hourly demand response considering intertemporal load characteristics," *IEEE Trans. Smart Grid*, vol. 2, no. 3, pp. 564–571, Sept. 2011.

33 J. Valenzuela and M. Mazumdar, "Statistical analysis of electric power production costs," *IIE Trans.*, vol. 32, pp. 1139–1148, 2000.

34 Available [online] https://rrtp.comed.com.

8

Microgrid Expansion Planning

8.1 Principles of Microgrid Expansion

Microgrids can eliminate investments in additional generation and transmission facilities to supply remote loads. Moreover, microgrid's islanding capability in the event of faults or disturbances in upstream networks would enhance grid and customers' reliability and resilience [1–12]. Microgrids can be categorized into different groups based on the type (such as campus, military, residential, commercial, and industrial), the size (such as small, medium, and large scales), the application (such as premium power, resilience-oriented, and loss reduction), and the connectivity (remote and grid-connected). Based on the voltages and currents adopted in a microgrid, however, three microgrid types can be identified: AC, DC, and hybrid. In AC microgrids, all distributed energy resources (DERs) and loads are connected to a common AC bus. DC generating units as well as energy storage will be connected to the AC bus via DC-to-AC inverters, and further, AC-to-DC rectifiers are used for supplying DC loads. In DC microgrids, however, the common bus is DC, where AC-to-DC rectifiers are used for connecting AC generating units, and DC-to-AC inverters are used for supplying AC loads. In hybrid microgrids, which can be considered as a combination of AC and DC microgrids, both types of buses exist, where the type of connection to each bus depends on the proximity of the DER/load to the bus. Extensive studies can be found on different aspects of microgrid operation and control, where the majority of these studies focus on AC microgrids, perceivably due to their connection to the AC utility grid and the utilization of AC DERs. DC microgrids can, however, offer several advantages when studied in detail and compared with AC microgrids, as follows: (i) higher efficiency and reduced losses due to the reduction of multiple converters used for DC loads; (ii) easier integration of various DC DERs, such as energy storage, solar photovoltaics (PV), and fuel cells, to the common bus with simplified

The Economics of Microgrids, First Edition. Amin Khodaei and Ali Arabnya.
© 2024 The Institute of Electrical and Electronics Engineers, Inc.
Published 2024 by John Wiley & Sons, Inc.

interfaces; (iii) more efficient supply of DC loads, such as electric vehicles and LED lights; (iv) eliminating the need for synchronizing generators, which enables rotary generating units to operate at their own optimum speed; and (v) enabling bus ties to be operated without the need for synchronizing the buses [13]. These benefits, combined with the significant increase in DC loads such as personal computers, laptop computers, LED lights, data and telecommunication centers, and other applications where the typical 50-Hz and 60-Hz AC systems are not available, can potentially introduce DC microgrids as a viable and economical solution in addressing future energy needs.

The prior research on DC microgrid planning is rather limited, and available studies on microgrid planning mostly focus on AC microgrids. The study in [14] proposes a planning model for AC microgrids considering uncertain physical and financial information. In this reference, the microgrid planning problem was broken down into an investment problem and an operation subproblem. The optimality of the solution was examined by employing the optimal planning decisions obtained from the master problem in the subproblem under uncertain conditions. The study in [15] suggests an operation modeling of hybrid AC–DC microgrids. It explains that the operation model of such a hybrid microgrid consists of system and device levels. This model includes the advantages of both AC and DC microgrids and performs both optimal scheduling and voltage control. The study in [16] proposes an operation planning model considering load/generation changes for a low-voltage DC microgrid, including DC sources such as batteries, fuel cells, and PVs. The objective of the study was to minimize daily operations costs. The model utilized a multi-path dynamic programming approach to solve the problem. The study in [17] presents multi-objective optimal scheduling of a DC microgrid consisting of a PV system and an electric vehicle charging station. In this reference, the cost of electricity and energy circulation in storage was taken as objective functions, and the mathematical model was built and solved to obtain the Pareto optimal solution. The study in [18] investigates a control system for hybrid AC–DC microgrids connected by multi-level inverters. The droop control technique was offered to manage power flows between AC microgrid, DC microgrid, and the main grid. The study in [19] discusses power management in a hybrid AC–DC microgrid and proposes an interlinking AC–DC converter accompanied by a suitable control system. The power flow between different sources throughout both microgrids was controlled in that study. The hybrid AC–DC microgrid allows different loads and DERs to connect with the minimum need for electrical conversion, which decreases the cost and energy losses. The study in [20] states that the efficiency of distributed generations and energy storage systems in a microgrid might reduce because of microgrid operation, hence running some consumers into a problem. This reference proposed an optimized operations planning for distributed generations and energy storage systems in microgrids to solve this issue.

In this chapter, it is assumed that the microgrid developer is planning to deploy a microgrid, however, the challenge is to determine the type of the microgrid between AC, DC, or hybrid, based on the system characteristics and accordingly determine the optimal DER generation mix. We propose a microgrid planning model with the overarching goals of (i) determining the optimal DER generation mix; (ii) determining the optimal type of the microgrid, i.e. either AC, DC, or hybrid, from an economic perspective; and (iii) identifying threshold ratios of DC loads, which make the DC microgrid a more economically viable alternative than the AC microgrid. The proposed microgrid planning model minimizes the total planning cost associated with the investment costs of DERs, AC-to-DC rectifiers, and DC-to-AC inverters, as well as the microgrid operation and reliability costs.

8.2 Economic Variables for Microgrid Expansion

8.2.1 AC vs. DC Microgrid Planning

The investment expenditure is typically higher for DERs compared to conventional energy resources within large-scale power plants due to the economies of scale of the latter. Nevertheless, DERs could provide less expensive energy in comparison with the energy purchased from the main grid, especially during peak hours when the market price is high. The energy storage could be used to be charged by the power from the main grid during low-price hours and discharged during high-price hours. One important and salient feature of microgrids that increases reliability is their islanding capability, which allows microgrids to be disconnected from the main grid in the presence of faults, disturbances, or voltage fluctuations in the upstream network. However, if after disconnecting from the main grid, the microgrid cannot supply all the loads, some loads should be curtailed, but critical loads will still be supplied. Another economic benefit of the microgrid is selling back the excess power to the main grid. The microgrid's economic viability is ensured when the total microgrid revenue from all available value streams in a specified time horizon exceeds the microgrid's total investment cost. The total planning cost comprised three parts, as follows: (i) the investment cost, (ii) the operations cost, and (iii) the reliability cost. The investment cost is long-term in nature and is calculated annually, while the operations and reliability costs are short-term and are calculated hourly for each day of the planning horizon.

In reality, several components should be considered to install the microgrid, but only the investment cost of DERs, rectifiers, and inverters is included at this point for the sake of simplicity of illustration of the model. Other costs associated with distribution network upgrades and installation of additional transformers, switches, measurement devices, and controllers are not considered since these costs will be similar in both types of microgrids. A general structure of DC microgrids is shown in Figure 8.1. In DC microgrids, three-phase AC-to-DC rectifiers and transformers are required to connect AC DERs to the common bus, single- and three-phase DC-to-AC inverters are needed for supplying AC loads, and a three-phase DC-to-AC/AC-to-DC converter, a transformer, and a point of common coupling (PCC) switch are required for connecting the microgrid to the utility grid.

A general structure of AC microgrids is shown in Figure 8.2. In AC microgrids, three-phase DC-to-AC inverters are required to connect DC DERs to the common bus, three-phase AC-to-DC rectifiers are needed for supplying DC loads, and similar to DC microgrids, a transformer and a PCC switch are required to connect the

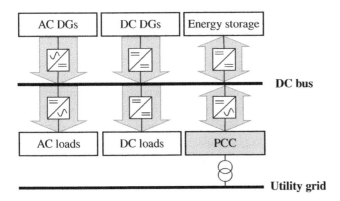

Figure 8.1 General structure of DC microgrids.

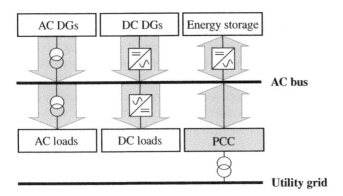

Figure 8.2 General structure of AC microgrids.

microgrid to the utility grid. The direction of arrows in Figures 8.1 and 8.2 shows the direction of power flow. It should be noted that different DC loads require different DC voltage levels, so some DC-to-DC converters must be considered as well in order to change the voltage level of the DC sources to desired levels. In both microgrids, a common bus is considered to show all the connections between loads and DERs. In practice, however, the common bus can represent one or more loop/radial distribution networks that connect loads and DERs within the microgrid. In DC microgrids, the common bus would handle DC voltages and currents, while in AC microgrids, the common bus would be used for AC voltages and currents.

The capacity of lines in a microgrid distribution network is typically much higher than the power transferred through the lines; therefore, the power flow is not considered in the proposed planning problem as the congestion is less likely and would not impact the planning results.

8.2.2 Hybrid Microgrid Planning

Figure 8.3 illustrates the general structure of a hybrid AC/DC microgrid. Without loss of generality, two AC and DC feeders are shown for presentation purposes. A PCC switch along with a transformer are required to connect both AC and DC buses to the utility grid. The AC bus can be directly connected to the PCC, while this connection for the DC bus can be made using a converter. A wide range of DERs can be used in the microgrid based on cost, availability, location, and system operator's preferences. However, the connection of each DER to its associated feeder needs to be made using proper converters. The DC DERs – such as solar PV, fuel cells, and distributed energy systems (DES), among others – need to be connected to AC feeders using DC-to-AC inverters. Similarly, AC DERs (such as wind turbine, co-gen, among others) need to be connected to the DC feeders using

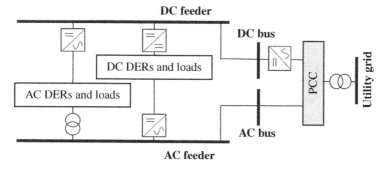

Figure 8.3 The general structure of a hybrid AC/DC microgrid.

AC-to-DC rectifiers. This is also the case for loads, in which the loads need to be connected to opposite-type feeders using proper converters.

We consider both renewable and dispatchable distributed generation units (DGs) for deployment in the microgrid. Renewable DGs have recently gained considerable traction, primarily due to the decreasing cost of renewable technology, economic incentives offered to customers, rapid construction and commissioning, and ease of installation compared to other types of DGs. Lacking, however, is the generation dispatchability that can ensure an economic and reliable supply of loads in grid-connected and islanded operation modes. Dispatchable DGs, therefore, are deployed in microgrids to ensure a controllable generation and guarantee an economic operation during grid-connected mode and an uninterrupted supply of critical loads during the islanded mode. The DES is further deployed to support renewable generation, enable energy arbitrage to increase economic benefits, and support microgrid control and islanding to increase reliability.

Without loss of generality, it is assumed that there are several feeders in the system, while the type of each feeder, i.e. AC or DC, needs to be determined. The decisive factors in determining the type of each feeder include the ratio of AC and DC loads at each feeder and the type of DERs connected to each feeder, which would accordingly impact the investment cost. Considering that the capacity of lines in a microgrid distribution network is typically much higher than the power transmitted through the lines, congestion would be less likely to occur. Therefore, the microgrid distribution network power flow and associated line limits are not considered in the proposed planning problem as they would not affect the planning results.

For the sake of illustration, in the rest of this chapter we focus on economic viability assessment modeling AC vs. DC microgrids, while interested readers are referred to [21] for modeling hybrid AC/DC microgrids.

8.3 Economic Viability Assessment Model

The objective of the microgrid planning problem is to minimize the microgrid total planning cost (8.1), which comprises the investment cost of DERs, rectifiers, and inverters (IC), the microgrid operations cost (OC), and the reliability cost (RC). The investment, operations, and reliability costs are determined in (8.2)–(8.5). Associated constraints are defined in (8.6)–(8.17). The type of the microgrid, i.e. either AC or DC, would impact the components to be installed in the microgrid, and accordingly, alter the investment cost. Constraints (8.2) and (8.3) define the DC investment cost and the AC investment cost, respectively, based on a binary decision variable z. If the microgrid is DC, the binary decision variable is set to one, relaxing

(8.3), and the investment cost would be determined by (8.2). Similarly, if the microgrid is AC, the binary decision variable is set to zero, relaxing (8.2), and the investment cost would be determined by (8.3).

$$\min\ IC + OC + RC \tag{8.1}$$

$$-M(1-z) \leq IC$$

$$-\left(\sum_t \sum_{i\in\{G,W\}} \kappa_t CC_{it} P_i^{\max} + \sum_t \sum_{i\in S} \kappa_t \left(CP_{it} P_i^{\max} + CE_{it} C_i^{\max}\right)\right.$$

$$\left.+\sum_t \sum_{i\in\{G_{ac},W_{ac}\}} \kappa_t CR_{it} P_i^{\max} + \sum_t \sum_{i\in I} \kappa_t CI_{it}(1-\alpha).\max(D_{bht}) + \sum_t \sum_{i\in I} \kappa_t CI_{it} P_M^{\max}\right)$$

$$\leq M(1-z) \tag{8.2}$$

$$-Mz \leq IC - \left(\sum_t \sum_{i\in\{G,W\}} \kappa_t CC_{it} P_i^{\max} + \sum_t \sum_{i\in S} \kappa_t \left(CP_{it} P_i^{\max} + CE_{it} C_i^{\max}\right)\right.$$

$$\left.+\sum_t \sum_{i\in\{G_{dc},W_{dc},S\}} \kappa_t CI_{it} P_i^{\max} + \sum_t \sum_{i\in R} \kappa_t CR_{it}\alpha.\max(D_{bht})\right) \leq Mz$$

$$\tag{8.3}$$

$$OC = \sum_t \sum_h \sum_b \sum_{i\in G} \kappa_t c_i P_{ibht} + \sum_t \sum_h \sum_b \kappa_t \rho_{bht} P_{M,bht} \tag{8.4}$$

$$RC = \sum_t \sum_h \sum_b \kappa_t \nu_{bht} LS_{bht} \tag{8.5}$$

where b is index for hour, h is index for day, i is index for DERs, t is index for year, G is set of all dispatchable units, G_{ac} is set of AC dispatchable units, G_{dc} is set of DC dispatchable units, I is set of DC-to-AC inverters, S is set of energy storage systems, W is set of all non-dispatchable units, W_{ac} is set of AC non-dispatchable units, W_{dc} is set of DC non-dispatchable units, c is generation price for dispatchable units, CC is annualized investment cost of generating units, CE is annualized investment cost of storage – energy, CI is annualized investment cost of DC-to-AC inverters, CP is annualized investment cost of storage – power, CR is annualized investment cost of AC-to-DC rectifiers, D is local demand, M is a large positive constant, $P_M^{\max}$ is flow limit between the microgrid and the main grid, α is ratio of DC loads to total loads, κ is coefficient of present-worth value, ρ is market price, ν is value of lost load (VOLL), $C^{\max}$ is variable for installed energy storage capacity, IC is variable for total investment cost, LS is variable for load curtailment, OC is variable for total operation cost, P is variable for DER output power, $P^{\max}$ is variable for installed

DER capacity, P_M is variable for main grid power, RC is variable for total reliability cost, and z is variable for microgrid investment state (0 if AC, 1 if DC). AC and DC microgrids have some similar components in the investment cost. The first two terms within the investment cost in (8.2) and (8.3) indicate the investment cost of DERs and energy storage, respectively. The investment cost of DERs depends on their installed power capacity, which will be determined by the optimization problem. The investment cost of energy storage further depends on its installed energy capacity. A single-step price curve is considered for DERs, which could be simply extended to a multi-step price curve. If the microgrid is DC, the output voltage of AC generating units should be converted to DC using rectifiers. Therefore, another term that should be considered is related to the investment cost of AC-to-DC rectifiers. Additionally, there are AC loads in the microgrids, requiring the use of DC-to-AC inverters. As a result, the investment cost of these inverters is included in the investment cost. The last term of the investment cost considers the DC-to-AC inverter, which is used for connecting the DC microgrid to the utility grid. For AC microgrids, as proposed in (8.3), DC-to-AC inverters must be used for connecting DC units to the microgrid, and AC-to-DC rectifiers are needed for supplying DC loads. These costs are included in the investment cost as well.

   The operations cost (8.4) includes the generation cost of dispatchable generating units and the cost of energy purchase from the main grid, which is defined as the amount of purchased energy times the market price at the PCC. If the microgrid is exporting its excess power to the main grid, the main grid power P_M would be negative (assumed to be paid at the market price under net metering); hence, there would be a benefit from selling the excess power. On the other hand, if there is a need for importing power from the main grid, P_M would be positive, increasing the operation costs. The reliability cost (8.5), which is the cost of unserved energy, is defined as the load curtailment quantity multiplied by the VOLL. VOLL represents customers' willingness to pay for reliable electricity service in order to avoid an outage. VOLL highly depends on the sector or customer type, timing of outage, duration of outage, and time of advanced notification of outage and preparation. Generally, VOLL for residential customers ranges from approximately $0/MWh to $17 976/MWh, while for commercial and industrial customers ranges from $3000/ MWh to $53,907/MWh [21]. Higher VOLLs represent more critical loads [22, 23]. A discount rate d is considered in order to evaluate the objective in terms of discounted costs. The present-worth cost component κ_t is present in all parts of the cost function and is calculated as $\kappa_t = 1/(1 + d)^{t-1}$. In (8.1)–(8.5), investment costs are calculated annually, while operations and reliability costs are calculated hourly and summed over all the years in the planning horizon.

   As described earlier, islanding is the most salient feature of microgrids, which enables the microgrid to be disconnected from the main grid in case of upstream network disturbances. In order to include the islanding ability of the microgrid, it is required to consider a condition to ensure that dispatchable generation capacity installed in the microgrid is adequate to seamlessly supply critical loads, as follows:

$$\beta \max(D_t) \leq \sum_{i \in G} P_i^{\max} \tag{8.6}$$

where parameter β represents the peak ratio of critical loads to total loads. The sum of the power from the main grid and from all DERs, including dispatchable and non-dispatchable units as well as energy storage, should be equal to the total load in each scheduling hour. Equations (8.7) and (8.8) consider the power balance equation in DC and AC microgrids, respectively. If the microgrid is DC, the binary decision variable is set to one, thus (8.8) would be relaxed and (8.7) would be applied. Similarly, if the microgrid is AC, (8.7) would be relaxed and (8.8) would be applied.

$$-M(1-z) \leq \left(\sum_{i \in \{G_{dc}, W_{dc}\}} P_{ibht} + \sum_{i \in S} \left(P_{ibht}^{dch} - P_{ibht}^{ch} \right) + \left(\sum_{i \in \{G_{ac}, W_{ac}\}} P_{ibht} + P_{M,bht} \right) \cdot \eta_{rec} \right.$$

$$\left. + LS_{bht} - \alpha \cdot D_{bht} - \frac{(1-\alpha) \cdot D_{bht}}{\eta_{inv}} \right) \leq M(1-z), \qquad \forall b, \forall h \tag{8.7}$$

$$-Mz \leq \left(\sum_{i \in \{G_{ac}, W_{ac}\}} P_{ibht} + \left(\sum_{i \in \{G_{dc}, W_{dc}\}} P_{ibht} + \sum_{i \in S} \left(P_{ibht}^{dch} - P_{ibht}^{ch} \right) \right) \cdot \eta_{inv} \right.$$

$$\left. + P_{M,bht} + LS_{bht} - (1-\alpha) \cdot D_{bht} - \frac{\alpha \cdot D_{bht}}{\eta_{rec}} \right) \leq Mz, \qquad \forall b, \forall h \tag{8.8}$$

where ch is superscript for energy storage charging mode, dch is superscript for energy storage discharging mode, *inv* is subscript for DC-to-AC inverters, *rec* is subscript for AC-to-DC rectifiers, and η is efficiency parameter (energy storage, inverters, and rectifiers). In DC microgrids, since power conversion causes power loss, an efficiency coefficient is defined in (8.7) for AC-to-DC rectifiers, used for converting the output of AC generating units and the power from the main grid, and for DC-to-AC inverters, used for supplying AC loads. Similar efficiency coefficients are considered for the AC microgrid (8.8).

The planning problem is further subject to constraints associated with the main grid power limits (8.9), dispatchable and non-dispatchable unit operations and planning ((8.10)–(8.12)), energy storage ((8.12)–(8.16)), and load curtailment (8.17), as follows:

$$-P_M^{\max} u_{bht} \leq P_{M,bht} \leq P_M^{\max} u_{bht} \qquad \forall b, \forall h \tag{8.9}$$

$$0 \le P_{ibht} \le P_i^{max} \qquad \forall i \in G, \forall b, \forall h \tag{8.10}$$

$$P_{ibht} = P_i^{max} \cdot pp_{ibht} \qquad \forall i \in W, \forall b, \forall h \tag{8.11}$$

$$P_i^{max} \le P_i^{cap} \qquad \forall i \in \{G, W, S\} \tag{8.12}$$

$$0 \le P_{ibht}^{dch} \le P_i^{max} \qquad \forall i \in S, \forall b, \forall h \tag{8.13}$$

$$0 \le P_{ibht}^{ch} \le P_i^{max} \qquad \forall i \in S, \forall b, \forall h \tag{8.14}$$

$$C_i^{max} \le C_i^{cap} \qquad \forall i \in S \tag{8.15}$$

$$0 \le \sum_{k \le b} \left(P_{ikht}^{ch} - P_{ikht}^{dch}/\eta_i \right) \le C_i^{max} \qquad \forall i \in S, \forall b, \forall h \tag{8.16}$$

$$0 \le LS_{bht} \le D_{bht} \qquad \forall b, \forall h \tag{8.17}$$

where C^{cap} is allowable energy storage installation capacity, P^{cap} is allowable DER installation capacity, pp is normalized forecast of non-dispatchable generation, and u is a binary islanding parameter. The amount of exchanged power with the main grid is limited by the capacity of the line connecting the main grid to the microgrid (8.9). In (8.9), the islanding capability of the microgrid is considered by defining a binary parameter that controls microgrid islanding. The power generated by dispatchable units is limited by their installed capacity (8.10). For non-dispatchable units, a variable and a parameter are used to consider their generation. Similar to dispatchable units, the variable P_i^{max} represents their installed capacity, which will be determined by solving the optimization problem. The parameter pp_{ibht} represents the normalized generation forecast of non-dispatchable units and has a value between 0 and 1 (8.11). Once a forecast is obtained, it is divided by the rated power of the candidate DER; hence, the normalized generation forecast is obtained. In this case, the selected size of the non-dispatchable unit will be considered as a scaling factor to scale up/down the normalized generation forecast and further obtain the actual generation. All DERs have an allowable installation capacity, and their installed capacity cannot exceed this limit (8.12). The allowable installation capacity may be obtained from budget limitations, choice of technology, or space limitations. The energy storage charging and discharging power in all hours is limited by its installed capacity ((8.13) and (8.14)). The installed energy capacity of the energy storage is limited by its allowable installation energy capacity (8.15). Additionally, its stored energy is determined based on the net charged power, efficiency, and stored energy in previous hours (8.16). It is further ensured that in case of local curtailments, the hourly curtailed load does not exceed the hourly total load (8.17).

8.4 Case Study

A microgrid is considered to be installed for a group of electricity customers with a peak annual load demand of 8.5 MW. The set of DERs used in this study includes four AC dispatchable units, one AC non-dispatchable unit (wind generator), one DC dispatchable unit (fuel cell), one DC non-dispatchable unit (solar PV), and one energy storage, as represented in Tables 8.1–8.3. The cost of converters is provided in Table 8.4. The load, renewable energy, and market price are forecasted based on

Table 8.1 Dispatchable units characteristics.

Unit No.	Type	Allowable installation capacity (MW)	Cost coefficient ($/MWh)	Annualized investment cost ($/MW)
1	Gas	5	90	50,000
2	Gas	5	90	50,000
3	Gas	3	70	70,000
4	Gas	3	70	70,000
5	Fuel cell	1.5	175	360,000

Table 8.2 Non-dispatchable units characteristics.

Unit No.	Type	Allowable installation capacity (MW)	Cost coefficient ($/MWh)	Annualized investment cost ($/MW)
6	Wind	2	0	132,000
7	Solar	2	0	133,000

Table 8.3 Energy storage characteristics.

Storage No.	Allowable installation capacity (MW)	Allowable installation energy (MWh)	Annualized investment cost – Power ($/MW)	Annualized investment cost – Energy ($/MWh)
1	1	6	60,000	30,000

Table 8.4 Annualized investment cost of converters.

Three-phase AC-to-DC rectifier ($/MW)	Single-phase DC-to-AC inverter ($/MW)	Three-phase DC-to-AC inverter ($/MW)
4200	6000	6500

historical data obtained from the Illinois Institute of Technology Campus Microgrid [24]. Data on wind, solar, fuel cells, and converters are gathered from [25–28]. The efficiency of energy storage and VOLL is considered to be 90% and $10,000/MWh, respectively. The planning horizon is 20 years. The lifetime of candidate DERs is equal to the planning horizon, i.e. 20 years. Twelve hours of islanding are considered in each planning year. The microgrid planning problem was implemented on a high-performance computing server consisting of four 10-core Intel Xeon E7-4870 2.4 GHz processors. The problem was formulated by mixed-integer programming (MIP) and solved by CPLEX 12.6 [29]. Following cases are studied.

Case 0: Baseline microgrid planning case;
Case 1: Sensitivity analysis on the ratio of DC loads;
Case 2: Sensitivity analysis on the ratio of critical loads;
Case 3: Sensitivity analysis on efficiency of AC-to-DC rectifiers and DC-to-AC inverters; and
Case 4: Sensitivity analysis on the market price.

Case 0: Initial values for the ratio of DC loads α, the ratio of critical loads β, and the efficiency of inverters and rectifiers η are chosen to be 0.40, 0.50, and 0.70, respectively. The microgrid planning solution would install dispatchable units 3 and 4 and the solar unit all with the maximum allowable capacity. The optimal planning solution would be the AC microgrid. The total planning cost in the base case is $25,608,640 with a cost breakdown of $6,679,653, $18,614,730, and $314,251 for the investment, operation, and reliability costs, respectively.

Case 1: In this case, the effect of changing the ratio of DC loads α on the type of the microgrid and installation of DERs is studied. The ratio of DC loads is changed by a step of 0.1, while all other parameters are kept unchanged. Results are represented in Tables 8.5 and 8.6. For values of α between 0 and 0.4, the optimal microgrid planning solution would install dispatchable units 3–4 and the solar unit, while by changing α between 0.5–0.8, dispatchable units 1 and 2 are also installed. However, for $\alpha = 0.9$ and 1, units 1 and 2 are not installed anymore, and the optimal microgrid planning solution would install the energy storage since the type of the microgrid is DC. The obtained results advocate that the installation of dispatchable units 3 and 4 with a higher investment cost is more economical than

Table 8.5 Installed DER capacity (MW) (β = 0.50, η = 0.70).

		DER							Storage	
α	z	1	2	3	4	5	6	7	P	E
0.00–0.40	0	0	0	3.0	3.0	0	0	2.0	0	0
0.50	0	0.03	0.03	3.0	3.0	0	0	2.0	0	0
0.60	0	0.15	0.15	3.0	3.0	0	0	2.0	0	0
0.70	0	0.27	0.27	3.0	3.0	0	0	2.0	0	0
0.80	0	0.40	0.40	3.0	3.0	0	0	2.0	0	0
0.90, 1.00	1	0	0	3.0	3.0	0	0	2.0	1.0	4.44

Table 8.6 Microgrid costs (β = 0.50, η = 0.70).

α	Investment cost	Operation cost	Reliability cost	Total cost
0.40	6,679,653	18,614,730	314,251	25,608,640
0.50	6,740,727	19,514,640	359,810	26,615,180
0.60	6,888,104	20,372,440	370,847	27,631,390
0.70	7,035,482	21,230,250	381,884	28,647,620
0.80	7,182,860	22,088,070	392,922	29,663,850
0.90	9,736,027	20,640,090	247,109	30,623,220
1.00	9,688,322	19,335,080	190,476	29,213,880

that of units 1 and 2. The reason is that units 3 and 4 offer a less expensive power compared to units 1 and 2. Additionally, between the two available non-dispatchable units, the solar unit is installed for all values of α although it has a higher investment cost than the wind unit since the generation pattern of the solar unit partially coincides with market price and load variations. The daily values of load, solar generation, and market price, averaged over one year, are shown in Figure 8.4 to demonstrate the partial correlation of the solar generation with the market price and the load. As shown in Figure 8.4, during the day, especially peak hours, the market price is higher, and the solar unit generates power. Therefore, part of the loads can be supplied by solar generation. On the other hand, wind energy is available mostly in early morning hours, when the market price is relatively low. As expected, according to the results and based on the values of β

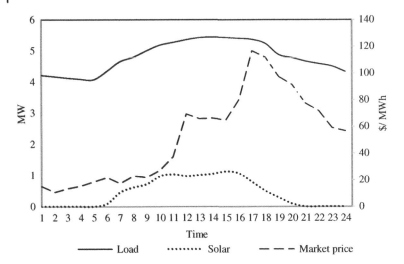

Figure 8.4 Annual average value of load and solar generation (MW) and the market price ($/MWh) for 24 hours.

and η, increasing the ratio of DC loads causes the microgrid to shift from AC (associated with $z = 0$) to DC (associated with $z = 1$). As shown in Table 8.6, by increasing α from 0.4 to 0.8, the microgrid investment cost increases because of the increased installed capacity of units 1 and 2 and the increased investment cost of rectifiers for supplying DC loads. For values of α between 0.4 and 0.8, the operations cost would increase as well since the amount of hourly power generated by dispatchable units 1 and 2 increases. By increasing α from 0.8 to 0.9, again the investment cost rises due to the installation of the energy storage, but the operations cost would decrease. The investment and operations costs would decrease by increasing α from 0.9 to 1. The investment cost drops as there are not any AC loads in the microgrid when $\alpha = 1$, thus the investment cost of inverters is eliminated. The operations cost drops as the overall exchanged power with the main grid decreases by changing all loads to DC. Accordingly, the microgrid's total planning cost would decrease by increasing α from 0.9 to 1. An interesting observation is the change in the total planning cost due to the change in the load mixture. According to results in Table 8.6, increasing the ratio of DC loads would cause an increase followed by a decrease in the total planning cost. Therefore, it would identify threshold ratios of DC loads, which make the DC microgrid economically a more viable solution than the AC microgrid. In other words, for ratios smaller than the threshold ratio, AC microgrid would be more economical, and for ratios larger than that, DC microgrid would be more economical.

Table 8.7 Installed DER capacity (MW) (α = 0.40, η = 0.70).

| | | | | DER | | | | | Storage | |
| | | | | | | | | | | |
β	z	1	2	3	4	5	6	7	P	E
0.10–0.70	0	0	0	3.0	3.0	0	0	2.0	0	0
0.80	0	0.40	0.40	3.0	3.0	0	0	2.0	0	0
0.90	0	0.82	0.82	3.0	3.0	0	0	2.0	0	0
1.00	0	1.25	1.25	3.0	3.0	0	0	2.0	0	0

Table 8.8 Microgrid costs (α = 0.40, η = 0.70).

β	Investment cost	Operation cost	Reliability cost	Total cost
0.10–0.70	6,679,653	18,614,730	314,251	25,608,640
0.80	7,050,504	18,433,520	165,911	25,649,930
0.90	7,448,045	18,238,630	76,485	25,763,160
1.00	7,845,585	18,043,630	0	25,889,220

Case 2: In this case, the effect of changing the ratio of critical loads β on planning results is investigated. Results are represented in Tables 8.7 and 8.8. The optimal microgrid planning solution would be the AC microgrid for all values of β. It is reasonable that by keeping α constant, there is no shift from the AC microgrid to the DC microgrid. The impact of β, however, can be noticed in the installed generation mix. According to Table 8.7, when the value of β is between 0.1 and 0.7, the microgrid planning solution would install dispatchable units 3 and 4 and the solar unit. By increasing the ratio of critical loads to 0.8 and more, units 1 and 2 are also installed, and their installed capacity would increase to supply critical loads. It is noticeable that the fuel cell, i.e. unit 5, would not be installed for any value of β. The reason is that the capital expenditure for the fuel cell is much higher than that of other DERs. It should be noted, however, that if the total critical load exceeds the total allowable DER capacity of available dispatchable units, the fuel cell would be installed as a last resort to ensure that critical loads would be supplied during islanding events. In other words, supply feasibility would become a more important factor than economic considerations.

Similar to Case 1, the solar unit is always installed due to the coincidence of its generation pattern with the load and market price variations. As shown in

Table 8.8, the operations and reliability costs would decrease by increasing β. Increasing the ratio of critical loads would cause an increase in the total installed DER capacity, while the total load has not changed. As a result, the excess power would be sold to the main grid, which increases the revenue of the microgrid thus decreasing the operations cost. On the other hand, by increasing the ratio of critical loads, the additional available dispatchable capacity would fully supply loads during islanding events, which causes load curtailments to decrease. Specifically, if all loads are considered as critical (associated with $\beta = 1.0$), the derived microgrid planning solution would install more dispatchable capacity so as to fully supply all loads, which causes load curtailments to reach zero at the expense of a higher investment cost.

Case 3: In this case, the effects of changing the efficiency of inverters and rectifiers η, which are considered to be equal, are investigated. Results show that changing converters' efficiencies while other parameters are kept unchanged would not affect either the type of microgrid or the installed DER mix. As shown in Table 8.9, the significant impact of changing η would be on the operations and reliability costs. By increasing η, there would be less power loss in inverters and rectifiers. Therefore, importing the power from the main grid during many operation hours would decrease, which would cause a reduction in the total operations cost. On the other hand, because of the reduced power loss in converters, more critical loads could be supplied by increasing efficiency. Accordingly, there would be a reduction in the load curtailment, which reduces the reliability cost. Since the installed power of all DERs is unchanged, the investment cost for different values of η would not change.

Case 4: In this case, the effect of changing the market price ρ on planning is studied. The installed power of DERs and costs associated with different market prices are represented in Tables 8.10 and 8.11, respectively. With a 10% decrease in the market price, the microgrid planning solution remains unchanged, except for the installed capacity of dispatchable units 3 and 4. Generally speaking, when the market price is low, the microgrid would buy more power from the main grid; hence, the exchanged power with the main grid would be positive in several hours.

Table 8.9 Microgrid costs ($\alpha = 0.40$, $\beta = 0.50$).

η	Investment cost	Operation cost	Reliability cost	Total cost
0.70	6,679,653	18,614,730	314,251	25,608,640
0.80		16,695,660	204,170	23,579,490
0.90		15,114,700	144,303	21,938,650
1.00		13,770,440	114,252	20,564,340

Table 8.10 Installed DER capacity (MW) (α = 0.40, β = 0.50, η = 0.70).

		DER								Storage	
Price change coefficient	z	1	2	3	4	5	6	7	P	E
0.9	0	0	0	2.91	2.91	0	0	2.0	0	0
Orig. price	0	0	0	3.00	3.00	0	0	2.0	0	0
1.1	0	1.23	1.23	3.00	3.00	0	0	2.0	0	0
1.2	0	5.00	5.00	3.00	3.00	0	2.0	2.0	0	0
1.3	0	5.00	5.00	3.00	3.00	0	2.0	2.0	0	0
1.4	0	5.00	5.00	3.00	3.00	0	2.0	2.0	0	0

Table 8.11 Microgrid costs (α = 0.40, β = 0.50, η = 0.70).

Price change coefficient	Investment cost	Operation cost	Reliability cost	Total cost
0.9	6,558,827	17,790,310	348,773	24,697,910
Orig. price	6,679,653	18,614,730	314,251	25,608,640
1.1	7,830,468	18,306,340	0	26,136,810
1.2	13,834,450	11,436,000	0	25,270,450
1.3	13,834,450	9,209,981	0	23,044,430
1.4	13,834,450	6,537,612	0	20,372,060

Therefore, the power generation of DERs would decrease in several hours, which leads to a reduction in the operations cost. Increasing the market price by 10% causes the optimal microgrid planning solution to install DERs 1 and 2 in addition to DERs 3, 4, and 7, thus the investment cost would increase. By increasing the market price by 20% or more, the microgrid should generate more power for several hours in order to supply loads, and on the other hand, it would be desirable to sell more electricity to the main grid. Therefore, all AC dispatchable units, wind generators, and solar PV would be installed at their maximum capacity, and the exchanged power with the main grid would be negative for several hours. As a result, the operations cost would decrease due to the revenue realized from selling more power to the main grid. It is further reasonable that all critical loads be supplied by increasing the total DER capacity. Accordingly, there would not be any load curtailment, which causes the reliability cost to reach zero. Since the DER generation mix is the same when there is a 20% or more increase in the market price, the investment cost would not change. Similar to previous cases, the type of the microgrid would remain the same, i.e. AC, since the ratio of DC loads is unchanged. It should be finally noted that since the capital investment cost of the fuel cell is too large, it would not be economical to install at any of the studied market prices.

Although in the proposed models it is assumed that annual changes in load, renewable generation, and market prices are negligible, the proposed microgrid planning model is capable of efficiently considering respective annual changes. Considering significantly small changes in the load is a practical assumption, perceivably due to the limited geographical boundaries of the microgrid, which limit significant load increase as well as the increased adoption of efficiency schemes, which helps with load reduction. In addition, renewable generation would remain the same over the planning horizon as the installed capacity would not change. The market price, however, has the highest possibility of increasing. To demonstrate the impact of the increase in the market price, the proposed planning problem is solved for a 2% annual increase in market prices. The total planning cost in this case is reduced to $24,635,350 with a cost breakdown of $9,296,503, $15,239,520, and $99,326 for the investment, operations, and reliability costs, respectively. Following the increase in market prices, the microgrid would be willing to sell more power to the main grid, which leads to a drop in the operations cost. On the other hand, in order to be able to sell more electricity, the microgrid would install additional DER capacity, which leads to an increase in the investment cost. The approximate computation time for each simulation was in the range of 118–155 minutes.

Arbitrary values for DERs' allowable installation capacity were used in these cases to show the effectiveness of the microgrid planning model in handling capacity limitations. If the limits are removed, the planning problem will select only the most economical candidates while ignoring all other candidates, which is not a very practical assumption. Some examples of these limitations are the rooftop solar panel installations in a community microgrid, which would be restricted by the rooftop area that can be covered by panels, and the thermal unit, which cannot be installed in densely populated areas.

These case studies demonstrate that DC microgrids can potentially improve the economic bottom line of microgrid projects when the ratio of DC loads is high, and further be considered as viable alternatives to AC microgrid installations. The takeaways from case studies are as follows:

- Among AC dispatchable generating units, those that offer less expensive power would be installed first, although they may be associated with higher capital costs.

- Among non-dispatchable units, the solar unit would be installed in all cases because of the partial coincidence of its generation pattern with the market price and load variations.
- The fuel cell would not be installed in any case since it is associated with a significantly higher investment cost compared to other DERs.
- The most decisive factor in determining the type of microgrid is the ratio of DC loads. Changing this ratio would cause the total cost to change, so it could be used as a tool to find a critical point where DC microgrid would be more economical than the AC microgrid.
- Increasing critical loads, converters' efficiency, or the market price would cause a decrease in the operation and reliability costs.
- An increase in critical loads would cause the optimal microgrid planning solution to install more dispatchable capacity, which increases the investment cost. Since the total load is unchanged, there would be excess generated power that would be sold to the main grid; hence, the operations cost would decrease. On the other hand, more critical loads would be supplied, which causes a decrease in the load curtailment and the reliability cost.
- Increasing converters' efficiency would cause a decrease in the power loss, which on one hand decreases the importing power from the main grid in many hours, thus decreasing the total operations cost, and on the other hand, more critical loads could be supplied; hence, there would be a reduction in the load curtailment, which reduces the reliability cost.
- The investment cost would change by changing the installed DER capacity. Therefore, the investment cost would remain unchanged by increasing η since the DER generation mix does not change.
- By increasing the market price, it would be desirable to install all dispatchable and non-dispatchable units, except for the fuel cell, in order to sell as much power as possible to the main grid, which would cause a decrease in the operation cost, and also supply all critical loads, thus decreasing the load curtailment, and accordingly, the reliability cost.

The proposed microgrid planning problem model could be further expanded to enhance practicality and computational efficiency. Specific areas for future work are identified as listed in the following.

8.4.1 Uncertainty Consideration

In the proposed model, forecasted data were used for hourly load, renewable energy, and market prices. Moreover, the islanding is considered within some specific hours in a planning year. Accurate data forecasting in microgrid planning models is a difficult task as there are various uncertainties in the planning data. In other words, there is an error associated with all forecasted values. Uncertainty considerations could potentially alter the microgrid planning results. This issue has been studied in the literature [14]. Similar methods can be applied here to expand the microgrid planning problem and make a more accurate decision between AC and DC microgrid installations.

8.4.2 Computational Complexity

The type of microgrid and DER generation mix in the proposed microgrid planning model are determined in an integrated fashion by solving a single optimization problem. This problem, however, is large-scale and nonconvex. A decomposition method can be employed in this case to convert the problem into a set of smaller and easier-to-solve, yet coordinated, subproblems. The application of decomposition methods in solving large-scale planning problems is extensively discussed in the literature and can be directly used here. A suggested decomposition for the proposed microgrid planning problem would include a long-term investment master problem, a short-term operation subproblem, and a reliability subproblem. The investment plan obtained in the master problem will be examined in subproblems to find the optimal DER schedule as well as desired levels of reliability. The final solution can be obtained in an iterative manner.

8.5 Summary

Among different categories of microgrids, i.e. AC, DC, and hybrid, extensive research has been conducted on operations and control of AC microgrids. DC microgrids can, however, offer several advantages compared to AC microgrids by providing a more efficient supply of DC loads, reducing losses due to the reduction of multiple converters used for DC loads, allowing easier integration of DC DERs, and eliminating the need for synchronizing generators. In this chapter, different components of AC, DC, and hybrid AC/DC microgrids were explained, followed by developing a microgrid planning model for determining the optimal size and the generation mix of DERs as well as the microgrid type, i.e. AC or DC. Considering the growing ratio of DC loads and DERs, DC microgrids can be potentially more beneficial than AC microgrids by avoiding the need to synchronize generators, reducing the use of converters, facilitating the connection of various types of DERs and loads to the microgrid common bus with simplified interfaces, and reducing losses associated with the AC–DC energy conversion. The problem objective was to minimize the total planning cost subject to prevailing planning and operational constraints and was formulated using MIP. The microgrid type is selected based on economic considerations, where the planning objective includes the investment and operations costs of DERs, cost of energy purchase from the main grid, and the reliability cost. It was shown that the proposed model could effectively identify threshold ratios of DC loads, which made the DC microgrid a more economically viable solution than the AC microgrid. In other words, for ratios smaller than the threshold ratio, AC microgrid would be more economical, and for ratios larger than that, DC microgrid would be more economical. A set of case studies was presented to analyze the impact of the ratio of DC loads, the ratio of critical loads, converters' efficiency, and market price on microgrid planning solutions. It was verified that the decisive factor in determining the type of microgrid would be the ratio of DC loads. In other words, if other parameters changed except for the ratio of DC loads, the type of the microgrid would not change. It was also shown that increasing the ratio of critical loads would increase the total installed dispatchable generation capacity. Finally, it was demonstrated that changing critical loads, converters' efficiency, or the market price would significantly affect the operations and reliability costs. Our case studies demonstrate the effectiveness of the proposed model and investigate in detail the impact of a variety of factors on planning results, including the ratio of critical loads, the ratio of DC loads, and the efficiency of inverters and converters.

References

1 M. Shahidehpour, "Role of smart microgrid in a perfect power system," in *IEEE Power Energy Soc. Gen. Meeting*, Minneapolis, MN, 2010.

2 M. Shahidehpour and J. Clair, "A functional microgrid for enhancing reliability, sustainability, and energy efficiency," *Electr. J.*, vol. 25, no. 8, pp. 21–28, Oct. 2012.

3 S. Bahramirad, W. Reder, and A. Khodaei, "Reliability-constrained optimal sizing of energy storage system in a microgrid," *IEEE Trans. Smart Grid*, vol. 3, no. 4, pp. 2056–2062, Dec. 2012.

4 N. Hatziargyriou, H. Asano, M. R. Iravani, and C. Marnay, "Microgrids: an overview of ongoing research, development and demonstration projects," *IEEE Power Energ. Mag.*, vol. 5, no. 4, July/Aug. 2007.

5 A. G. Tsikalakis and N. D. Hatziargyriou, "Centralized control for optimizing microgrids operation," *IEEE Trans. Energy Convers.*, vol. 23, no. 1, Mar. 2008.

6 B. Kroposki, R. Lasseter, T. Ise, S. Morozumi, S. Papathanassiou, and N. Hatziargyriou, "Making microgrids work," *IEEE Power Energ. Mag.*, vol. 6, no. 3, May 2008.

7 S. Parhizi, H. Lotfi, A. Khodaei, and S. Bahramirad, "State of the art in research on microgrids: a review," *IEEE Access*, vol. 3, pp. 890–925, July 2015.

8 A. Khodaei, "Microgrid optimal scheduling with multi-period islanding constraints," *IEEE Trans. Power Syst.*, vol. 29, no. 3, pp. 1383–1392, May 2014.

9 A. Khodaei, "Resiliency-oriented microgrid optimal scheduling," *IEEE Trans. Smart Grid*, vol. 5, no. 4, pp. 1584–1591, July 2014.

10 A. Khodaei, and M. Shahidehpour, "Microgrid-based co-optimization of generation and transmission planning in power systems," *IEEE Trans. Power Syst.*, vol. 28, no. 2, pp. 1582–1590, May 2013.

11 S. Bahramirad, A. Khodaei, J. Svachula, and J. R. Aguero, "Building resilient integrated grids: one neighborhood at a time," *IEEE Electrificat. Mag.*, vol. 3, no. 1, pp. 48–55, Mar. 2015.

12 H. Nikkhajoei and R. H. Lasseter, "Distributed generation interface to the CERTS microgrid," *IEEE Trans. Power Delivery*, vol. 24, no. 3, pp. 1598–1608, July 2009.

13 P. Cairoli and R. A. Dougal, "New horizons in DC shipboard power systems: new fault protection strategies are essential to the adoption of dc power systems," *IEEE Electrificat. Mag.*, vol. 1, no. 2, pp. 38–45, Dec. 2013.

14 A. Khodaei, S. Bahramirad, and M. Shahidehpour, "Microgrid planning under uncertainty," *IEEE Trans. Power Syst.*, no. 99, pp. 1–9, Oct. 2014.

15 P. Teimourzadeh Baboli, M. Shahparasti, M. Parsa Moghaddam, M. R. Haghifam, and M. Mohamadian, "Energy management and operation modelling of hybrid AC–DC microgrid," *IET Gener. Transm. Distrib.*, vol. 8, no. 10, pp. 1700–1711, Oct. 2014.

16 C. Huang, M. Chen, Y. Liao, and C. Lu, "DC microgrid operation planning, renewable energy research and applications (ICRERA)," in *2012 International Conference*, pp. 1–7, 2012.

17 X. Lu, N. Liu, Q. Chen, and J. Zhang, "Multi-objective optimal scheduling of a DC micro-grid consisted of PV system and EV charging station," in *2014 IEEE Innovative Smart Grid Technologies – Asia (ISGT Asia)*, Kuala Lumpur, Malaysia, pp. 487–491, 2014.

18 V. K. Hema and R. Dhanalakshmi, "Analysis of power sharing on hybrid AC-DC microgrid," in *2014 Annual International Conference on Emerging Research Areas: Magnetics, Machines and Drives (AICERA/iCMMD)*, Kottayam, India, pp. 1–6, 2014.

19 N. Eghtedarpour and E. Farjah, "Power control and canagement in a hybrid AC/DC microgrid," *IEEE Trans. Smart Grid*, vol. 5, no. 3, pp. 1494–1505, May 2014.

20 E. Shimoda, S. Numata, J. Baba, T. Nitta, and E. Masada, "Operation planning and load prediction for microgrid using thermal demand estimation," in *2012 IEEE Power and Energy Society General Meeting*, San Diego, CA, pp. 1–7, 2012.

30 H. Lotfi and A. Khodaei, "Hybrid AC/DC microgrid planning," *Energy*, vol. 118, pp. 37–46, Jan. 2017.

21 Estimating the Value of Lost Load [Online]. Available: http://www.ercot.com/content/gridinfo/resource/2015/mktanalysis/ERCOT_ValueofLostLoad_LiteratureReviewandMacroeconomic.pdf.

22 C. K. Woo, and R. L. Pupp, "Cost of service disruptions to electricity consumers," *Int. J. Energy*, vol. 17, no. 2, pp. 109–126, Feb. 1992.

23 Y. L. Mok and T. S. Chung, "Prediction of domestic, industrial and commercial interruption costs by relational approach," in *Proc. 4th Int. Conf. Advances in Power System Control, Operation and Management*, Hong Kong, vol. 1, pp. 209–215, 1997.

24 Illinois Institute of Technology Campus Data [Online]. Available: http://www. iitmicrogrid.net/microgrid.aspx.

25 S. Schoenung, "Energy Storage Systems Cost Update," SAND2011-2730, 2011.

26 Distributed Generation Renewable Energy Estimate of Costs [Online]. Available: http://www.nrel.gov/analysis/tech_lcoe_re_cost_est.html.

27 R. Remick and D. Wheeler, "Molten Carbonate and Phosphoric Acid Stationary Fuel Cells: Overview and Gap Analysis," [Online]. Available: http://www.nrel.gov/docs/fy10osti/49072.pdf, 2010.

28 List of Converters Prices [Online]. Available: http://www.alibaba.com/product-detail/2wm-3wm-5wm-rectifier-system-and_60148157747.html.

29 CPLEX 12, IBM ILOG CPLEX, User's Manual, [Online]. Available: http://gams.com/dd/docs/solvers/cplex.pdf, 2013.

9

Microgrids for Asset Management in Power Systems

9.1 Principles of Asset Management

Asset management denotes management and engineering practices applied to valuable assets of a system in order to deliver the required level of service to the customers. Asset management is a critical responsibility of electric utility companies to maintain power network reliability and quality of service at acceptable levels by reducing the failure probability of critical grid components. In other words, asset management extends the remaining useful life (RUL) of equipment and decreases the risk of equipment failure and unplanned power outages. Considering that the majority of current power grid is aging as they were mainly built several decades ago and at the same time the customers' expectations for a high quality of service is at all-time high, the topic of asset management is more important than ever [1–4].

Transformers are one of the most critical electrical equipment particularly when it comes to asset management, primarily due to their impact on power system adequacy and reliability. Transformer failures can potentially lead to unplanned power outages, in addition to costly and time-consuming repairs and replacement [2–4]. Condition monitoring, online monitoring, routine diagnostic, scheduled maintenance, and condition-based maintenance (CBM) are some of the most common transformer asset management methods [2, 5, 6]. The RUL of a transformer highly depends on its insulation condition owing to a higher probability of insulation failure compared with its other components. Moreover, the aging of transformer insulation is a function of insulation moisture, oxygen amount, and internal temperature specifically at the hottest spot, which is mainly governed by transformer loading and ambient temperature [7–9]. In [4], power transformer asset management is performed using a two-stage maintenance scheduler. The effect of temperature, thermal aging factors, and electrical aging factors on transformer insulation are experimentally analyzed in [10]. In [11], an experimental thermal model for 25 kVA transformers is proposed which estimates the transformer's RUL and accordingly the time of transformer maintenance or replacement.

The Economics of Microgrids, First Edition. Amin Khodaei and Ali Arabnya.
© 2024 The Institute of Electrical and Electronics Engineers, Inc.
Published 2024 by John Wiley & Sons, Inc.

A method for calculating transformer insulation loss of useful life is provided as a standard, IEEE Std. C57.91-2011 Guide for Loading Mineral-Oil-Immersed Transformers, in [12]. Authors in [13] present a sensory model framework in which the transformer's RUL is estimated based on the measured values of winding hottest-spot temperature and the aforementioned IEEE standard. The study in [7] proposes a model for estimating the RUL of transformer insulation via this IEEE standard, based on historical data of load and ambient temperature. A fuzzy modeling in [14] is applied for transformer asset management for improvement in RUL of transformers. Application of different machine learning methods, such as Adaptive Network-Based Fuzzy Inference System (ANFIS), Multi-Layer Perceptron (MLP) network and Radial Basis Function (RBF) network, in estimating transformer loss of useful life is presented in [15], while these methods were fused together to improve the estimation accuracy in [16]. In [17], an artificial neural network is modeled to predict top oil temperature in a transformer, where ambient temperature and load current are considered as the input layer and top oil temperature as the output layer. Since transformer loading has the most significant effect on transformer insulation loss of useful life, its management and control can remarkably increase transformer lifetime. In [8, 18, 19], the effect of electric vehicles on distribution net load profile and accordingly on distribution equipment such as transformers is studied, and a smart charging method is proposed to manage distribution and transmission assets, including transformers, via controlling and managing distribution net load profile. The effect of electric vehicles and rooftop solar photovoltaic on distribution transformer aging is investigated in [20, 21]. These studies show that rooftop solar generation decreases transformer loss of useful life, as it reduces the power transferred from the utility grid to loads, while electric vehicles reduce the RUL of transformers, and their charging/discharging should be controlled to prevent negative impacts on the connected transformer's RUL. A control algorithm with the objective of controlling the electric load of plug-in electric vehicle on distribution transformer is proposed in [22]. The proposed algorithm aims at reducing distribution transformer overloading via leveraging vehicle-to-grid strategy. An electric vehicle charging algorithm is studied in [23] to coordinate the grid and distribution transformer. The algorithm can prevent the distribution transformer from overloading and sharp ramping through smoothing the transformer load profile.

In this chapter, a method for distribution transformer asset management by leveraging microgrids is proposed. Considering the wide range of benefits that microgrids can provide to the power systems [24–35], in this chapter, we aim building up on existing research and deployment efforts and focuses on the flexibility advantages of the utility-owned microgrids as a complementary value proposition for distribution transformer asset management. The microgrid capability in managing its adjustable loads, dispatchable Distributed Generation

(DG) units, Distributed Energy Storage (DES) units, and the ability of exchanging power with the utility grid in the grid-connected mode is specifically considered in this chapter for smoothing distribution transformer loading, and consequently decreasing transformer loss of useful life which leads to higher transformer's RUL. It is assumed that the microgrid is utility-owned, thus can be scheduled by the electric utility company or any designated entity as the operator.

By leveraging the IEEE Std. C57.91-2011, the distribution transformer loss of useful life is calculated to be integrated in the microgrid optimal scheduling model. The aforementioned standard for calculation of the distribution transformer loss of useful life has a set of nonlinear equations which would make the microgrid optimal scheduling a nonlinear and hard to solve problem. To ensure that the microgrid optimal scheduling problem keeps its linear characteristics, the original problem is decomposed into a master problem as a mixed-integer linear program (MILP) minimizing the microgrid operation cost, and a nonlinear subproblem that determines the distribution transformer loss of useful life using Benders' decomposition. These two problems are further coordinated through Benders cuts in an iterative manner. Using this proposed iterative method, the master problem solves the microgrid optimal scheduling problem, as discussed in several existing research work such as [28–30], while the added subproblem provides feedback on how microgrid operations would impact the transformer's RUL, and accordingly, would provide a signal (the Benders cut) on how microgrid schedule should change to increase transformer's RUL. It should be noted that although the proposed models are based on the IEEE Std. C57.91-2011, without loss of generality, any other standard or updates to this standard can be modeled using the same approach in the proposed model.

9.2 Economic Variables in Microgrid Asset Management

9.2.1 The IEEE Standard – Guide for Loading Mineral-Oil-Immersed Transformers

The IEEE Std.C57.91-2011 proposes a set of nonlinear functions to calculate transformer loss of useful life. Equation (9.1) formulates aging acceleration factor (F_{AA}) for a given load and ambient temperature, where θ^H is a function of transformer load profile and ambient temperature:

$$F_{AA} = \exp\left(\frac{15,000}{383} - \frac{15,000}{\theta^H + 273}\right). \tag{9.1}$$

Equation (9.1) is used in calculating the equivalent aging of the transformer in a desired time interval as in (9.2), which can be considered as one unit of time (can be hour, day, month, etc.)

$$F_{EQA} = \sum_{n=1}^{N} F_{AA_n} \Delta t_n \Big/ \sum_{n=1}^{N} \Delta t_n, \tag{9.2}$$

where Δt_n is time interval, n is the time interval index and N is the total number of time intervals. Accordingly, the percentage of insulation loss of life (LOL) is calculated, as follows:

$$LOL(\%) = \frac{F_{EQA} \times t \times 100}{\text{Normal insulation life}}. \tag{9.3}$$

Based on the IEEE Std. C57.91-2011 [12], normal RUL for insulation of distribution transformers is 180,000 hours. As it can be seen in (9.1), θ^H is the backbone term to calculate transformer loss of useful life. Based on (9.4), the hottest-spot temperature is composed of three distinct terms:

$$\theta^H = \theta^A + \Delta\theta^{TO} + \Delta\theta^H, \tag{9.4}$$

where, θ^A represents ambient temperature, $\Delta\theta^{TO}$ is top-oil rise over ambient temperature, and $\Delta\theta^H$ is the winding hottest-spot rise over top-oil temperature. $\Delta\theta^{TO}$ and $\Delta\theta^H$ are defined in (9.5) and (9.6), respectively, as follows:

$$\Delta\theta^{TO} = \left(\Delta\theta^{TO,U} - \Delta\theta^{TO,I}\right)\left(1 - \exp\left(-\frac{1}{\tau^{TO}}\right)\right) + \Delta\theta^{TO,I} \tag{9.5}$$

$$\Delta\theta^H = \left(\Delta\theta^{H,U} - \Delta\theta^{H,I}\right)\left(1 - \exp\left(-\frac{1}{\tau^{W}}\right)\right) + \Delta\theta^{H,I} \tag{9.6}$$

Moreover, (9.7)–(9.10) calculate the initial and the ultimate values for $\Delta\theta^{TO}$ and $\Delta\theta^H$:

$$\Delta\theta^{TO,I} = \Delta\theta^{TO,R}\left[\frac{\left(K^I\right)^2 R + 1}{R + 1}\right]^n, \tag{9.7}$$

$$\Delta\theta^{TO,U} = \Delta\theta^{TO,R}\left[\frac{\left(K^U\right)^2 R + 1}{R + 1}\right]^n, \tag{9.8}$$

$$\Delta\theta^{H,I} = \Delta\theta^{H,R}\left(K^I\right)^{2m}, \tag{9.9}$$

$$\Delta\theta^{H,U} = \Delta\theta^{H,R}\left(K^U\right)^{2m}, \tag{9.10}$$

where K^I and K^U are respectively initial and ultimate values of transformer load ratio at each time interval. Both m and n can vary between 0.8 and 1 based on the transformer cooling mode [12].

Taking aforementioned equations into account, it can be seen that the percentage value for loss of useful life at each time interval is a nonlinear function of initial/ultimate values of transformer load ratio, and ambient temperature, i.e. K^I, K^U and θ^A, respectively. In other words, by knowing K^I, K^U and θ^A at each time interval, the percentage value for loss of useful life can be calculated via the sequence of these nonlinear functions. One key point is that θ^A can be forecasted accurately for each location at each time interval so that the percentage value for loss of useful life, as defined in (9.11), will be a nonlinear function of initial and ultimate values of transformer load ratio, i.e. K^I and K^U, respectively:

$$LOL(\%) = f\left(K^I_{ht}, K^U_{ht}\right) \qquad \forall h, \forall t. \tag{9.11}$$

9.2.2 Transformer Asset Management via Microgrid Optimal Scheduling

The proposed extended microgrid optimal scheduling problem determines the least-cost schedule of available resources – including distributed energy resources (DERs) and loads – while minimizing the cost of distribution transformer loss of useful life (9.12), subject to prevailing operational constraints (9.13)–(9.39), as follows:

$$\min \sum_h \sum_t \left[\sum_{i \in G} F_i(P_{iht}) + \rho^M_{ht} P^M_{ht} \right] + \psi f\left(K^I_{ht}, K^U_{ht}\right) \tag{9.12}$$

s.t.

$$\sum_i P_{iht} + P^M_{ht} = \sum_d D_{dht} \qquad \forall h, \forall t, \tag{9.13}$$

$$-P^{M, \max} w_{ht} \leq P^M_{ht} \leq P^{M, \max} w_{ht} \qquad \forall h, \forall t, \tag{9.14}$$

$$P^{\min}_i I_{iht} \leq P_{iht} \leq P^{\max}_i I_{iht} \qquad \forall i \in G, \forall h, \forall t, \tag{9.15}$$

$$P_{iht} - P_{ih(t-1)} \leq UR_i \qquad \forall i \in G, \forall h, t \neq 1, \tag{9.16}$$

$$P_{ih1} - P_{i(h-1)T} \leq UR_i \qquad \forall i \in G, \forall h, \forall t, \tag{9.17}$$

$$P_{ih(t-1)} - P_{iht} \leq DR_i \qquad \forall i \in G, \forall h, t \neq 1, \tag{9.18}$$

$$P_{i(h-1)T} - P_{ih1} \leq DR_i \qquad \forall i \in G, \forall h, \forall t, \tag{9.19}$$

$$T_i^{\text{on}} \geq UT_i\left(I_{iht} - I_{ih(t-1)}\right) \qquad \forall i \in G, \forall h, t \neq 1, \tag{9.20}$$

$$T_i^{\text{on}} \geq UT_i\left(I_{ih1} - I_{i(h-1)T}\right) \qquad \forall i \in G, \forall h, \forall t, \tag{9.21}$$

$$T_i^{\text{off}} \geq DT_i\left(I_{ih(t-1)} - I_{iht}\right) \qquad \forall i \in G, \forall h, t \neq 1, \tag{9.22}$$

$$T_i^{\text{off}} \geq DT_i\left(I_{i(h-1)T} - I_{ih1}\right) \qquad \forall i \in G, \forall h, \forall t, \tag{9.23}$$

$$P_{iht} \leq P_{iht}^{\text{dch, max}} u_{iht} - P_{iht}^{\text{ch, min}} v_{iht} \qquad \forall i \in S, \forall h, \forall t, \tag{9.24}$$

$$P_{iht} \geq P_{iht}^{\text{dch, min}} u_{iht} - P_{iht}^{\text{ch, max}} v_{iht} \qquad \forall i \in S, \forall h, \forall t, \tag{9.25}$$

$$u_{iht} + v_{iht} \leq 1 \qquad \forall i \in S, \forall h, \forall t, \tag{9.26}$$

$$C_{iht} = C_{ih(t-1)} - \left(P_{ith} u_{iht} \tau^{ES} / \eta_i\right) - P_{iht} v_{iht} \tau^{ES} \qquad \forall i \in S, \forall h, t \neq 1, \tag{9.27}$$

$$C_{ih1} = C_{i(h-1)T} - \left(P_{ih1} u_{iht} \tau^{ES} / \eta_i\right) - P_{ih1} v_{iht} \tau^{ES} \qquad \forall i \in S, \forall h, \forall t, \tag{9.28}$$

$$C_i^{\text{min}} \leq C_{iht} \leq C_i^{\text{max}} \qquad \forall i \in S, \forall h, \forall t, \tag{9.29}$$

$$T_{iht}^{\text{ch}} \geq MC_i\left(u_{iht} - u_{ih(t-1)}\right) \qquad \forall i \in S, \forall h, t \neq 1, \tag{9.30}$$

$$T_{ih1}^{\text{ch}} \geq MC_i\left(u_{ih1} - u_{i(h-1)T}\right) \qquad \forall i \in S, \forall h, \forall t, \tag{9.31}$$

$$T_{iht}^{\text{dch}} \geq MD_i\left(v_{iht} - v_{ih(t-1)}\right) \qquad \forall i \in S, \forall h, t \neq 1, \tag{9.32}$$

$$T_{ih1}^{\text{dch}} \geq MD_i\left(v_{ih1} - v_{i(h-1)T}\right) \qquad \forall i \in S, \forall h, \forall t, \tag{9.33}$$

$$D_d^{\text{min}} z_{dht} \leq D_{dht} \leq D_d^{\text{max}} z_{dht} \qquad \forall d \in D, \forall h, \forall t, \tag{9.34}$$

$$T_d^{\text{on}} \geq MU_d\left(z_{dht} - z_{dh(t-1)}\right) \qquad \forall d \in D, \forall h, t \neq 1, \tag{9.35}$$

$$T_d^{\text{on}} \geq MU_d\left(z_{dh1} - z_{d(h-1)T}\right) \qquad \forall d \in D, \forall h, \forall t, \tag{9.36}$$

$$\sum_{[\alpha, \beta]} D_{dht} = E_d \qquad \forall d \in D, \tag{9.37}$$

$$\left(\left|\hat{P}_{ht}^M\right| / P_{\text{nom}}^{\text{Trans}}\right) = K_{ht}^U \qquad \forall h, \forall t, \tag{9.38}$$

$$\left(\left|\hat{P}_{h(t-1)}^M\right| / P_{\text{nom}}^{\text{Trans}}\right) = K_{ht}^I \qquad \forall h, \forall t. \tag{9.39}$$

where d is index for loads, h is index for day, i is index for DERs, t is index for hour, ch is superscript for distributed energy storage charge, dch is superscript for distributed energy storage discharge, H is superscript for winding hottest-spot, I is superscript for initial value of variables and parameters, R is superscript for rated value, TO is superscript for transformer top oil, U is superscript for ultimate value of variables and parameters, W is transformer winding, $\wedge$ is superscript for calculated/given variables, D is set of adjustable loads, G is set of dispatchable units, S is set of energy storage units, DR is ramp down rate parameter, DT is minimum downtime parameter, E is load total required energy parameter, $F(.)$ is generation cost parameter, F_{AA} is aging acceleration factor of insulation parameter, MC is minimum charging time parameter, MD is minimum discharging time, MU is minimum operating time parameter, m/n is an empirically derived exponent to calculate the variation of $\Delta\theta^{H}/\Delta\theta^{TO}$ with changes in load, R is ratio of full-load loss to no-load loss, Δt is time interval, UR is ramp up rate, UT is minimum uptime, w is binary islanding indicator (1 if grid-connected, 0 if islanded), α, β are specified start and end times of adjustable loads, ρ is market price, η is energy storage efficiency, ψ is transformer investment cost, τ is time period, θ is temperature (°C), C is variable for energy storage available (stored) energy, D is variable for load demand, I is variable for commitment state of dispatchable units, K is variable for transformer loading ratio, P is variable for DER output power, P^{M} is variable for utility grid power exchange with the microgrid, P^{M1}, P^{M2} are slack variables for utility grid power, Q is variable for the cost of transformer loss of useful life, T^{ch} is variable for number of successive charging hours, T^{dch} is variable for number of successive discharging hours, T^{on} is variable for number of successive ON hours, T^{off} is variable for number of successive OFF hours, u is variable for energy storage discharging state (1 when discharging, 0 otherwise), v is variable for energy storage charging state (1 when charging, 0 otherwise), z is variable for adjustable load state (1 when operating, 0 otherwise).

The first term in the objective function (9.12) minimizes the microgrid annual operations cost, including the local generation cost and the cost of energy exchange with the utility grid. The second term represents the cost of distribution transformer loss of useful life. This term consists of multiplication of distribution transformer's loss of useful life, based on the IEEE Standard explained in Section 9.2, and the distribution transformer investment cost (ψ). This term attempts to minimize the distribution transformer loading in order to reduce its loss of useful life and consequently increase its RUL. This investment cost is used to ensure that both terms in the objective have a similar unit (here $). It should also be noted that the maintenance cost of generation units has been already included in the first term of the objective function (9.12) as the local generation cost. The load balance Eq. (9.13) ensures that the summation of power exchange with the utility grid and the local generations (including dispatchable DGs, non-dispatchable DGs, and the DES) would be equal to microgrid total load at each operating hour. The DES power can be positive (discharging), negative (charging) or zero (idle). In addition, the power exchange between microgrid and the utility grid (P^{M}) could be positive (import), negative (export) or zero. This power is also restricted to the capacity of the line between the microgrid and the utility grid (9.14). Hourly generation of dispatchable DGs are constrained by the maximum and minimum capacity limits (9.15), where the unit commitment state variable I would be 1 when the unit is committed and 0 otherwise. Constraints (9.16)–(9.19) represent ramp up and ramp down constraints of dispatchable DG units, where (9.16) and (9.18) belong to intra-day intervals and (9.17) and (9.19) represent ramping constraints for inter-day intervals. Dispatchable DG units are subject to the minimum uptime and downtime limits, represented by (9.20)–(9.23). Constraints (9.20), (9.22) and (9.21), (9.23) represent minimum uptime/downtime for inter-day and intra-day intervals, respectively. Constraints (9.24) and (9.25) respectively define the minimum and maximum limits of the DES charging and discharging. It should be noted that in the charging/discharging mode the binary charging/discharging state variable v/u is 1/0 and the binary discharging/charging state variable u/v is 0/1. Constraint (9.26) ensures that the DES can merely operate in one mode of charging or discharging at any given period. The amount of charged and discharged power in the DES and the available stored energy would determine the stored energy in intra-day (9.27) and inter-day (9.28) intervals, where one hour is considered for time period of charging and discharging. The amount of stored energy in DES is further limited to its capacity (9.29). Constraints (9.30), (9.32) and (9.31), (9.33) represent the minimum charging/discharging times of DES for intra-day and inter-day intervals, respectively. Constraint (9.34) confines adjustable loads to minimum and maximum rated powers, while (9.35) and (9.36) represent the minimum operating time of adjustable loads for intra-day and inter-day intervals. It should be noted that in (9.34)–(9.36), when load is on, binary operating variable z is 1, otherwise it is 0. Moreover, (9.37) considers the required energy to complete an operating cycle for adjustable loads. Note that the adjustable loads utilized here are responsive to price changes and controlling signals from the microgrid controller so that no compensation costs are considered. It should be mentioned that $b = 0$, which would appear in (9.17), (9.19), (9.21), (9.23), (9.28), (9.31), (9.33), (9.36), represents the last day of the previous scheduling horizon, and T represents the last scheduling hour, i.e. $T = 24$.

As the exchanged power between the microgrid and the utility grid (P^M) determines the distribution transformer load ratio, i.e. K^U and K^I, constraints (9.38) and (9.39) are needed to represent the interdependency of these variables. Based on the direction of power exchange between the microgrid and the utility grid, the amount of P^M can be positive (exporting power) or negative (importing power), but the transformer load ratio (K^I or K^U) can only take positive values. Thus, the absolute value of P^M should be considered in (9.38) and (9.39), which represent the relationship between the transformer loading and the microgrid power exchange with the utility grid.

9.3 An Economic Model for Integration of Microgrids in Asset Management

Figure 9.1 depicts the flowchart of the proposed microgrid-based distribution transformer asset management model using Benders' decomposition method. The objective of the original microgrid-based distribution transformer asset management model is the summation of microgrid operations cost and the distribution transformer's cost of loss of useful life, i.e. the summation of a linear and a nonlinear term. However, in Benders' decomposition the subproblem does not need to be necessarily in a linear form [36]. We use Benders' decomposition to decompose the microgrid-based distribution transformer asset management problem to a mixed integer linear programming master problem (minimizing the microgrid operation cost) and a nonlinear subproblem (determines the distribution transformer loss of useful life). These two problems are further coordinated through optimality cuts in an iterative manner. Using this proposed iterative method,

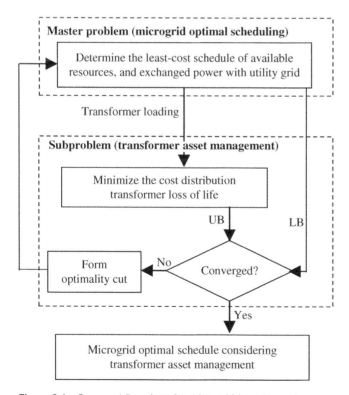

Figure 9.1 Proposed flowchart for microgrid-based transformer asset management.

the master problem solves the microgrid optimal scheduling problem, while the added subproblem acts as feedback on how microgrid operations can impact the transformer's RUL, and accordingly, would provide a signal (the optimality cut) on how microgrid schedule should change to increase transformer's RUL. The step-by-step procedure for microgrid-based distribution transformer asset management solution is, as follows:

Step 1: Solve the microgrid optimal scheduling master problem by considering the commitment and dispatch of available DGs, the charging and discharging schedules of DESs, the schedule of adjustable loads, and the exchanged power with the utility grid. Note that there is no optimality cut available in the first iteration of the master problem.

Step 2: Minimize the transformer asset management subproblem by considering the exchanged power of the microgrid with the utility grid (transformer loading).

Step 3: Compare the subproblem's solution, i.e. an upper bound, with the solution of the master problem, i.e. a lower bound. If the difference is larger than a predetermined threshold, form the optimality cut and send back to the master problem to consequently revise the current schedule of available resources and the exchanged power with the utility grid. Otherwise, consider the microgrid-based distribution transformer asset management solution as optimal.

Branch-and-bound is a commonly used technique for solving mixed integer linear programming problems. Two processes are employed in this technique as follows: (i) bounding process; and (ii) branching process. In the bounding process, the solution of a relaxed mixed integer linear programming problem, i.e. converting mixed integer linear programming problem into linear programming problem via removing integrity restrictions, is calculated and then imposed as a lower bound for minimization problems or upper bound for maximization problems. In the branching process, the problem is broken into two subproblems, where further are solved to obtain the solutions. If the solutions for both of these subproblems satisfy the integrity conditions, they are compared with each other, and the subproblem solution related to smaller objective function value for minimization problem or larger one for maximization problem will be selected as the optimal solution. Note that if only one of these two subproblems solution satisfies the mixed integer linear program's integrity condition, this solution is kept as an incumbent solution (i.e. the optimal solution if no better solution will be achieved further). Nevertheless, the branching process is continued to search on the other subproblem with the objective of finding a better solution that satisfied the mixed integer linear programming integrity condition [37]. Mixed integer linear program solvers such as CPLEX, Xpress-MP, SYMPHONEY, and CBC, reap the benefits of a combination of branch-and-bound techniques and cutting-plane techniques to accelerate the computation time associated with solving mixed integer linear programming problems, which consequently facilitate solving large mixed integer linear programming problems using personal computers. The branch-and-bound technique for solving mixed integer nonlinear programs is based on the same idea as the branch-and-bound technique employed to solve MILPs. Similar to the branch-and-bound technique explained above, the technique starts by solving the problem where the discrete conditions of the binary variables are relaxed. If the obtained solution is integral, then this solution is considered as an optimal solution for the problem. Without loss of generality, the two processes of bounding and branching are employed in order to find the optimal solution for the mixed integer nonlinear programming problem [38]. The optimality of the Benders' decomposition method is extensively discussed in the following references [36, 39, 40].

9.3.1 Microgrid Optimal Scheduling (Master Problem)

The objective of the microgrid optimal scheduling master problem is to minimize the microgrid annual operations cost, subject to (9.13)–(9.37). The second term added to the objective function is the projected cost of the distribution transformer loss of useful life, which will be obtained from the optimality cuts generated in the transformer asset management subproblem. The value of this term in the first iteration will be 0. The master problem determines the optimal microgrid schedule, where the optimal values of the exchanged power between the microgrid and the utility grid will be sent to the distribution asset management subproblem with the objective of calculating the optimal value for the distribution transformer loss of useful life.

$$\min \sum_{h} \sum_{t} \left[\sum_{i \in G} F_i(P_{iht}) + \rho_{ht}^M P_{ht}^M \right] + \Lambda, \tag{9.40}$$

Subject to (9.13)–(9.37), where Λ is variable for reflected cost for transformer loss of useful life in the master problem.

9.3.2 Transformer Asset Management (Subproblem)

The objective of the transformer asset management subproblem is to minimize the cost of distribution transformer loss of useful life based on the IEEE Std. C57.91-2011, as defined in (9.41), and subject to additional limitations on the distribution transformer loading (9.38) and (9.39).

$$\min Q = \sum_{h} \sum_{t} \psi f\left(K_{ht}^I, K_{ht}^U\right) \tag{9.41}$$

$$P^M_{h(t-1)} = \hat{P}^M_{h(t-1)} \qquad \lambda_{ht} \qquad \forall h, \forall t, \tag{9.42}$$

$$P^M_{ht} = \hat{P}^M_{ht} \qquad \mu_{ht} \qquad \forall h, \forall t. \tag{9.43}$$

The exchanged power of the microgrid with the utility grid (transformer loading) is calculated in the master problem and used in the subproblem as given values in (9.42) and (9.43). λ_{ht} and μ_{ht} are dual variables associated with the initial and ultimate microgrid exchanged power with the utility grid at each time interval, respectively. These dual variables are calculated through linearization of subproblem around the operating point in each iteration, determined in the master problem.

The solution of the original integrated problem based on the current obtained solution would provide an upper bound (9.44), while the lower bound in each iteration is the solution of the master problem, i.e. microgrid annual operation cost plus the term reflecting cost of transformer loss of useful life.

$$UB = \sum_h \sum_t \left[\sum_{i \in G} F_i\left(\hat{P}_{iht}\right) + \rho^M_{ht} \hat{P}^M_{ht} \right] + \psi f\left(\hat{K}^I_{ht}, \hat{K}^U_{ht}\right). \tag{9.44}$$

The final solution of the original problem is achieved when the difference between these two bounds is smaller than a threshold. If the convergence criterion is not satisfied, the optimality cut (9.45), is generated and added to the master problem to revise the solution in the next iteration.

$$\Lambda \geq \hat{Q} + \sum_h \sum_t \lambda_{ht} \left(\left| P^M_{h(t-1)} \right| - \left| \hat{P}^M_{h(t-1)} \right| \right) + \sum_h \sum_t \mu_{ht} \left(\left| P^M_{ht} \right| - \left| \hat{P}^M_{ht} \right| \right), \tag{9.45}$$

where $\hat{Q}$ is the calculated objective value for the distribution transformer's loss of useful life (optimal solution for the subproblem). Moreover, the optimality cut (9.45) consists of two terms associated with the initial and ultimate microgrid exchanged power with the utility grid. This cut indicates that the solution of the revised microgrid optimal scheduling could lead to a better solution for the transformer asset management subproblem, i.e. the one which causes a smaller cost for the distribution transformer loss of useful life. The absolute function in (9.45) makes the master problem nonlinear. In order to have a linear model in the master problem, two new nonnegative variables (P^{M1} and P^{M2}) are considered in a way that only one of them can be selected via binary variables x and y (9.46) and (9.47). As P^{M1}, P^{M2} are both nonnegative variables and only one of them can be nonzero at every hour, in case of power export ($P^M > 0$) $P^M = P^{M1}$ and $P^{M2} = 0$, and similarly, in case of power import ($P^M < 0$) $P^M = -P^{M2}$ and $P^{M1} = 0$.

$$P_{ht}^M = x_{ht}P_{ht}^{M1} - y_{ht}P_{ht}^{M2} \qquad \forall h, \forall t, \tag{9.46}$$

$$x_{ht} + y_{ht} \le 1 \qquad \forall h, \forall t. \tag{9.47}$$

where x, y are binary variables for selecting slack variables associated with utility grid exchange, Multiplication of binary variables (x and y) with continues variables (P^{M1} and P^{M2}) makes bilinear terms ($x_{ht}P^{M1}$ and $y_{ht}P^{M2}$) in (9.46), which are linearized via (9.48)–(9.50), with M as a large positive constant.

$$-Mx_{ht} - My_{ht} \le P_{ht}^M \le Mx_{ht} + My_{ht} \qquad \forall h, \forall t, \tag{9.48}$$

$$P_{ht}^{M1} - M(1 - x_{ht}) \le P_{ht}^M \le P_{ht}^{M1} + M(1 - x_{ht}) \qquad \forall h, \forall t, \tag{9.49}$$

$$-P_{ht}^{M2} - M(1 - y_{ht}) \le P_{ht}^M \le -P_{ht}^{M2} + M(1 - y_{ht}) \qquad \forall h, \forall t. \tag{9.50}$$

If binary variables x and y are zero, P^M would be 0 and (9.49) and (9.50) would be relaxed. If binary variables x or y are 1, (9.48) would be relaxed and P^M would be equal to either P^{M1} or $-P^{M2}$, based on (9.49) and (9.50), respectively. In order to have a positive value for P^M in (9.45), this variable is replaced with the summation of P^{M1} and P^{M2} which leads to a revised representation of the optimality cut, as follows:

$$\begin{aligned}
\Lambda \ge \hat{Q} + \sum_h \sum_t \lambda_{ht} \left[\left(P_{h(t-1)}^{M1} + P_{h(t-1)}^{M2} \right) - \left(\hat{P}_{h(t-1)}^{M1} + \hat{P}_{h(t-1)}^{M2} \right) \right] \\
+ \sum_h \sum_t \mu_{ht} \left[\left(P_{ht}^{M1} + P_{ht}^{M2} \right) - \left(\hat{P}_{ht}^{M1} + \hat{P}_{ht}^{M2} \right) \right]
\end{aligned} \tag{9.51}$$

The optimality cut (9.45) plays a key role for restricting the lower bound of the microgrid optimal scheduling master problem. Using the proposed Benders' decomposition procedure in an iterative manner between the master problem and the subproblem, a decomposed model for the microgrid-based distribution transformer asset management will be achieved. This model benefits from reshaping microgrid exchanged power with the utility grid to maximize the distribution transformer's RUL.

9.4 Case Study

To investigate the performance of the proposed model, a test microgrid system which consists of four dispatchable DGs, two non-dispatchable DGs (G5: wind and G6: solar), one DES, and five adjustable loads is considered and studied. The characteristics of generation units, energy storage system, and adjustable loads are tabulated in Tables 9.1–9.3, respectively. The forecasted values for microgrid hourly fixed load, non-dispatchable units' generation, and market price for

Table 9.1 Characteristics of generation units.

Unit	Type	Cost coefficient ($/MWh)	Min–max capacity (MW)	Min up/down time (h)	Ramp up/down rate (MW/h)
G1	D	27.7	1–5	3	2.5
G2	D	39.1	1–5	3	2.5
G3	D	61.3	0.8–3	1	3
G4	D	65.6	0.8–3	1	3
G5	ND	0	0–1	—	—
G6	ND	0	0–1.5	—	—

D, dispatchable; ND, non-dispatchable.

Table 9.2 Characteristics of the energy storage system.

Storage	Capacity (MWh)	Min–max charging/discharging power (MW)	Min charging/ discharging time (h)
ESS	10	0.4–2	5

Table 9.3 Characteristics of adjustable loads.

Load	Type	Min–max capacity (MW)	Required energy (MWh)	Initial start-end time (h)	Min up time (h)
L1	S	0–0.4	1.6	11–15	1
L2	S	0–0.4	1.6	15–19	1
L3	S	0.02–0.8	2.4	16–18	1
L4	S	0.02–0.8	2.4	14–22	1
L5	C	1.8–2	47	1–24	24

S, shiftable; C, curtailable.

one sample day are provided in Tables 9.4–9.6, respectively. Note that scheduling horizon of one year is considered in this chapter. More details on the hourly loads and market price for the considered one-year operation are available in [41]. A 10 MVA distribution transformer is considered at the Point of Common Coupling (PCC) with the characteristics borrowed from [7]. The nominal active power of

Table 9.4 Microgrid hourly fixed load (one day as a sample).

Time (h)	1	2	3	4	5	6
Load (MW)	8.73	8.54	8.47	9.03	8.79	8.81
Time (h)	7	8	9	10	11	12
Price ($/MWh)	10.12	10.93	11.19	11.78	12.08	12.13
Time (h)	13	14	15	16	17	18
Price ($/MWh)	13.92	15.27	15.36	15.69	16.13	16.14
Time (h)	19	20	21	22	23	24
Price ($/MWh)	15.56	15.51	14.00	13.03	9.82	9.45

Table 9.5 Generation of non-dispatchable units (one day as a sample).

Time (h)	1	2	3	4	5	6
G5 (MW)	0	0	0	0	0.63	0.80
G6 (MW)	0	0	0	0	0	0
Time (h)	7	8	9	10	11	12
G5 (MW)	0.62	0.71	0.68	0.35	0.62	0.36
G6 (MW)	0	0	0	0	0	0.75
Time (h)	13	14	15	16	17	18
G5 (MW)	0.4	0.37	0	0	0.05	0.04
G6 (MW)	0.81	1.20	1.23	1.28	1.00	0.78
Time (h)	19	20	21	22	23	24
G5 (MW)	0	0	0.57	0.60	0	0
G6 (MW)	0.71	0.92	0	0	0	0

Table 9.6 Hourly electricity price (one day as a sample).

Time (h)	1	2	3	4	5	6
Price ($/MWh)	15.03	10.97	13.51	15.36	18.51	21.8
Time (h)	7	8	9	10	11	12
Price ($/MWh)	17.3	22.83	21.84	27.09	37.06	68.95
Time (h)	13	14	15	16	17	18
Price ($/MWh)	65.79	66.57	65.44	79.79	115.45	110.28
Time (h)	19	20	21	22	23	24
Price ($/MWh)	96.05	90.53	77.38	70.95	59.42	56.68

the distribution transformer is considered to be 10 MW. In order to calculate the transformer loss of useful life, the hourly forecasted ambient temperature of a specific location in Houston, Texas [42] for one year is used. Since this study does not take into account power congestion and power flow calculations, the system topology diagram is not of significance and the results are independent of the topology.

In order to investigate the effectiveness of the proposed model, the six cases are studied, as follows. Case 0 represents transformer's loss of useful life calculation. Case 1 represents microgrid optimal scheduling ignoring transformer asset management constraints. Case 2 represents microgrid optimal scheduling considering transformer asset management constraints. Case 3 represents microgrid optimal scheduling with limited transformer overloading while ignoring asset management constraints. Case 4 represents microgrid optimal scheduling with limited transformer overloading and asset management constraints. Finally, Case 5 represents sensitivity analysis with regards to market price forecast errors, transformer loading, and adjustable loads.

Case 0: In this case, it is assumed that the microgrid loads are only supplied by the utility grid, i.e. the local generation is ignored. The transformer loading in this case is similar to the microgrid load profile, as the exchanged power with the utility grid to supply the microgrid load passes through the transformer. The annual transformer loss of useful life in this case is calculated as 3.1%, which represents an expected RUL of 32 years.

Case 1: The grid-connected price-based optimal scheduling is analyzed for a one-year horizon. In the price-based scheduling the main goal is to minimize the microgrid operations cost without any commitments in supporting transformer asset management. The microgrid operation cost is calculated as $1,632,296, and the annual transformer loss of useful life is calculated as 2.7% in this case. If this value is considered as the average annual loss of useful life, an expected lifetime of 37 years is perceived for the transformer. The primary reason for this longer RUL (37 years) compared to the value calculated in Case 0 (32 years) is the microgrid local generation which would partially supply local loads and thus reduce the transformer loading. This situation leads to a smaller loss of useful life and consequently longer RUL for the distribution transformer. In other words, even without considering asset management in microgrid scheduling, the transformer's RUL will be prolonged as the microgrid reduces transformer loading through local generation and partial load offset. It should however be noted that possible transformer overloading is ignored in this case.

Case 2: In this case, the microgrid controller minimizes the microgrid operations cost while considering the transformer asset management constraints. In other words, in addition to minimizing the operation cost, the microgrid controller attempts to reduce the transformer loading which leads to lowering the transformer loss of useful life, and consequently translates into longer RUL. The annual transformer loss of useful life is reduced from 2.7% in Case 1 to 2.08%, at the expense of a 0.11% increase in microgrid operations cost compared to Case 1 to reach a cost of $1,634,239. The transformer's RUL is increased in this case by an average of 11 years. Two points can be considered, as follows. (i) this considerable increase in the transformer's RUL is achieved by the insignificant addition of less than $2000/year to the microgrid operation cost; and (ii) transformer is not overloaded in any of the operation hours, i.e. the microgrid only reshapes the transformer loading profile without causing any overloads. The considerable impact of overloads will be further discussed in following cases.

Figure 9.2 compares the exchanged power with utility grid in Cases 1 and 2 in one day, as a sample from the one-year optimal scheduling horizon. As shown in Figure 9.2, when the mere aim of the microgrid in Case 1 is minimizing its operations cost, the power is purchased from the utility grid when the market price is low, and the extra power is sold back to the utility grid when the market price is high. In other words, the economic incentive is the only major factor in determining the optimal schedule. However, in Case 2, in addition to optimal scheduling of the microgrid, the distribution transformer loss of useful life is considered, so the exchanged power is reshaped in order to reduce load variations. Explicitly, power exchange changed in hours 13, 15, and 18 as it is more economical to reduce the transformer loading rather than purchasing less expensive power from or selling extra power to the utility grid.

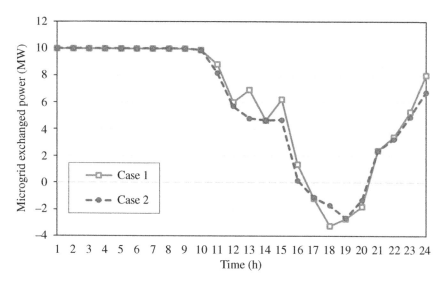

Figure 9.2 Microgrid exchanged power with the utility grid in Cases 1 and 2 in a sample day of the studied year.

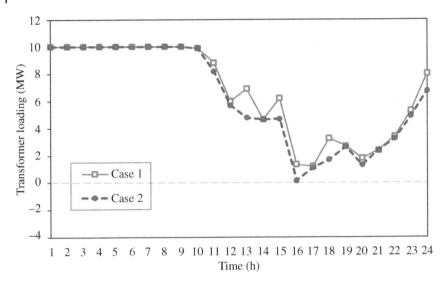

Figure 9.3 Transformer loading in Cases 1 and 2 in a sample day of the studied year.

Figure 9.3 depicts the transformer loading in both cases in the same studied day, which better illustrates the effect of the transformer asset management constraints on the microgrid power exchange. The depicted transformer loading is the absolute value of microgrid exchanged power with the utility grid shown in Figure 9.2. As shown, the transformer loading is reduced in the range between 0.1 MW (at hour 17) and 2.1 MW (at hour 13). This decrease causes a reduction in the transformer's loss of useful life on this specific day from 0.0040% to 0.00367%. This reduced rate is the effect of applying transformer asset management in the microgrid optimal scheduling during only one sample day of the studied year.

Case 3: The transformer overloading is considered in this case, without taking the transformer asset management constraints into account. A 20% overloading at 3 hours (13, 14, and 15) of 20 random days in a year is considered, that is only 60 hours of 8760 hours in a year. Figure 9.4 shows the transformer loading in this case and compares it with that of Case 1 (without transformer overloading). As Figure 9.4 shows, a three-hour overloading in the afternoon not only leads to changes in the transformer loading pattern during the transformer overloaded hours, but also impacts the transformer loading in the remaining hours of the studied day. The transformer's loss of useful life in this case is increased to 3.09% compared to 2.7% in the case without overloads.

The results show that the initial transformer loss of useful life of 0.0065% is increased to 0.0264% in this sample day only due to a three-hour overload.

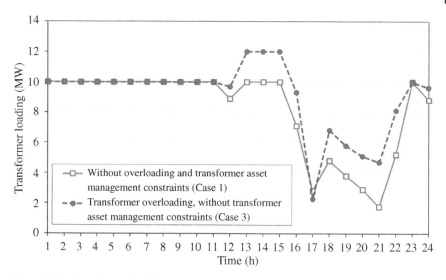

Figure 9.4 Transformer loading in Case 3 in one of the days with transformer overloading as a sample.

This significant rise of the transformer's loss of useful life (more than four times) shows the considerable effects of the transformer overloading on its RUL reduction. This increase occurs due to the exponential nature of the equations used in calculating the transformer loss of useful life. The microgrid operations cost in this case is calculated as $1,628,345. It should be noted that the sample day in this case, shown in Figures 9.4 and 9.5, is selected from the 20 studied days for transformer overloading, and it is not the same as the selected day in Figures 9.2 and 9.3.

Case 4: The parameters and conditions of this case are similar to those in Case 3, while the transformer asset management constraints are considered, as well. By adding the transformer asset management constraints, as Figure 9.5 demonstrates, the transformer loading decreases not only during the overloading hours but also during most hours after the overloading. The changes in microgrid schedule and energy arbitrage lead to 22% decrease in the transformer loss of useful life (2.41% in this case compared with 3.09% in Case 3). However, this drop of the transformer loss of useful life and increasing its RUL led to a higher microgrid operations cost, calculated as $1,630,842 in this case.

The obtained results of the studied cases are shown in Table 9.7. As the results from Cases 0 and 1 demonstrate, utilizing a microgrid significantly decreases the annual transformer loss of useful life and consequently increases the expected RUL of the transformer. A comparison between Cases 1 and 2 advocates that taking transformer asset management constraints into account leads to decreasing the

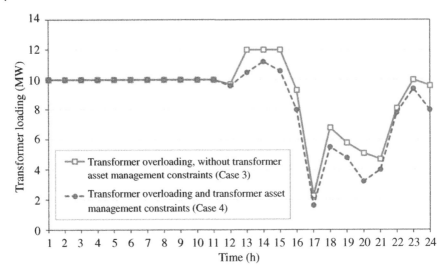

Figure 9.5 Comparison of transformer loading in Cases 3 and 4, transformer overloading with and without transformer asset management.

Table 9.7 Microgrid operation cost and transformer loss of useful life and RUL for studied cases.

Cases	Annual microgrid operation cost ($)	Annual transformer loss of useful life (%)	Transformer expected RUL (years)
Case 0	–	3.1	32
Case 1	1,632,296	2.7	37
Case 2	1,634,239	2.08	48
Case 3	1,628,345	3.09	32.3
Case 4	1,630,842	2.41	41.5

annual transformer loss of useful life even further (48 years in Case 2, compared to 37 years in Case 1), while the annual microgrid operations cost marginally increases. A comparison between Cases 3 and 4 also highlights the impact of the transformer asset management constraints on reducing the transformer loss of useful life under transformer overloading conditions.

Case 5: The sensitivity of derived results with regards to market price prediction errors, transformer loading, and adjustable loads are thoroughly investigated in this case, as follows.

Case 5(a) – Sensitivity analysis with regards to market price prediction errors: A sensitivity analysis is performed to study the impact of prediction errors on annual transformer loss of useful life, transformer expected RUL, and annual microgrid operations cost. Prediction errors of $\pm 10\%$, $\pm 20\%$, and $\pm 30\%$ are considered for the annual hourly market price. The derived results for this sensitivity analysis are shown in Table 9.8. As shown, the annual transformer loss of useful life drops by increasing market price prediction errors, and accordingly the transformer expected RUL increases. When market price increases, the master controller readjusts the microgrid schedule with the objective of supplying the loads locally rather than importing power from the main grid. Nevertheless, the microgrid exchanged power with the utility grid, i.e. transformer loading, is decreased which translates into a lower transformer loss of useful life and a higher transformer expected RUL, in cases of ignoring and considering transformer asset management constraints. In addition, the results demonstrate that the annual transformer loss of useful life as well as the transformer expected RUL are significantly improved by taking the transformer asset management constraints into account. For instance, in case of "30% decrease" and "30% increase," the transformer expected RUL increases by 6 years and 12.5 years, respectively.

It should be noted that the annual microgrid operations cost slightly rises by considering transformer asset management constraints, at the expense of lowering the transformer loss of useful life and increasing the transformer expected RUL.

Case 5(b) – Sensitivity analysis with regards to transformer loading: The effects of transformer loading on the annual transformer loss of useful life as well as the transformer expected RUL are investigated in this case. To this end, 50%, 75%, 100%, and 125% of the transformer nominal power (P_n) are considered as the maximum limitation for the transformer loading. The obtained results from this case study are shown in Table 9.9. The sensitivity results clearly show the exponential increase in transformer's loss of useful life by increasing the transformer loading. By keeping the transformer loading within the limit of 50%, the annual transformer loss of useful life is calculated respectively as 0.455% and 0.452% in case of ignoring and considering transformer asset management constraints. On the other hand, overloading the distribution transformer will dramatically reduce its RUL. The transformer loss of useful life under 125% transformer loading, i.e. 25% overload, is respectively calculated as 11.83% and 8.61% in cases of ignoring and considering transformer asset management constraints – where the transformer expected RUL will be 8.5 and 11.6 years, respectively. Moreover, the results demonstrate that the transformer expected RUL will be increased slightly while taking the transformer asset management constraints into account for lower

Table 9.8 Sensitivity analysis with regards to market price forecast error.

	Annual transformer loss of useful life (%)		Transformer expected RUL (years)		Annual microgrid operations cost ($)	
Market price	Ignoring transformer asset management constraints	Considering transformer asset management constraints	Ignoring transformer asset management constraints	Considering transformer asset management constraints	Ignoring transformer asset management constraints	Considering transformer asset management constraints
30% decrease	3.41	2.83	29.3	35.3	1,242,627	1,245,215
20% decrease	3.077	2.41	32.5	41.5	1,396,111	1,399,733
10% decrease	2.84	2.23	35.2	44.8	1,525,675	1,528,842
Default	2.7	2.08	37.0	48.1	1,632,296	1,634,239
10% increase	2.57	2.011	38.9	49.7	1,715,356	1,717,944
20% increase	2.51	1.935	39.8	51.7	1,776,963	1,779,887
30% increase	2.456	1.88	40.7	53.2	1,821,077	1,823,412

Table 9.9 Sensitivity analysis with regards to transformer loading.

Transformer loading (%)	Annual transformer loss of useful life (%)		Transformer expected RUL (years)	
	Ignoring transformer asset management constraints	Considering transformer asset management constraints	Ignoring transformer asset management constraints	Considering transformer asset management constraints
50	0.455	0.452	219.8	221.2
75	1.67	1.38	59.9	72.5
100	2.7	2.08	37	48.1
125	11.83	8.61	8.5	11.6

transformer loading limits. It should be noted that the cases with very low/high limits of transformer loading, i.e. 50% or 125%, are not practical and just are considered in this study as extreme operating conditions.

Case 5(c) – Sensitivity analysis with regards to adjustable loads: To demonstrate the effect of adjustable loads on the annual transformer loss of useful life, transformer's expected RUL, and annual microgrid operations cost, the problem is solved for various cases of adjustable loads. The required energy of the five aggregated adjustable loads is changed from 10 to 50 MWh (which can be considered as having more adjustable loads in the microgrid). The obtained results for this study are provided in Table 9.10. As the sensitivity analysis results show, by increasing the adjustable loads, the annual transformer loss of useful life is slightly reduced, which means the transformer expected RUL increases. By changing the total required energy of adjustable loads from 10 to 50 MWh, the transformer expected RUL increases by 1.2 years from 48.05 to 49.25 years, when taking the transformer asset management constraints into account. In addition, as the total required energy of adjustable loads increases, the annual microgrid operations cost reduces in both cases of ignoring and considering transformer asset management constraints. Nevertheless, adjustable loads play a key role in reshaping the loading of the distribution transformer at the point of interconnection in order to increase its RUL. The cost associated with the power loss is significantly smaller than the transformer loss of useful life and microgrid operations costs so that its impacts will be negligible. Nevertheless, in order to ensure this assumption, a case study is performed in which 6% distribution power loss is considered in the distribution deployed microgrid. The obtained results demonstrate that the cost associated with the power loss is a very small fraction of the transformer loss of useful life and microgrid operations costs. Therefore, if the power loss cost of the microgrid is taken into consideration, the results will be affected to a minimal extent; however, the final assessment and the conclusion remain intact.

Table 9.10 Sensitivity analysis with regards to adjustable load.

Adjustable load	Annual transformer loss of useful life (%)		Transformer expected RUL (years)		Annual microgrid operations cost ($)	
	Ignoring transformer asset management constraints	Considering transformer asset management constraints	Ignoring transformer asset management constraints	Considering transformer asset management constraints	Ignoring transformer asset management constraints	Considering transformer asset management constraints
Default	2.704	2.081	36.98	48.05	1,632,296	1,634,239
10 MWh increase	2.698	2.071	37.07	48.30	1,590,523	1,590,890
20 MWh increase	2.684	2.060	37.25	48.55	1,553,959	1,557,723
30 MWh increase	2.672	2.050	37.43	48.78	1,520,413	1,524,829
40 MWh increase	2.663	2.041	37.56	49.00	1,496,362	1,499,589
50 MWh increase	2.650	2.030	37.73	49.25	1,476,587	1,478,510

9.5 Summary

Asset management is an important function performed by utility companies to increase the RUL of their critical power infrastructure assets to ensure grid reliability by minimizing unexpected outages. This chapter introduced a microgrid application for distribution asset management as a viable and cost-effective alternative to traditional utility practices in this area. A microgrid is an emerging distribution technology that encompasses a variety of distribution technologies including distributed generation, demand response, and energy storage. Moreover, the substation transformer, as the most critical component in a distribution grid, is selected as the component of choice for asset management studies. We proposed a microgrid-based distribution transformer asset management model. In the proposed model, microgrid exchanges power with the utility grid in such a way that the distribution transformer's RUL is maximized. Using Benders' decomposition method, the proposed model was decomposed into a microgrid optimal scheduling master problem and a distribution transformer asset management subproblem. Based on a relevant IEEE standard, the optimal cost of the distribution transformer loss of useful life was calculated in the subproblem in order to examine the optimality of the microgrid scheduling solution. This means that the distribution transformer's asset management subproblem was presented to manipulate the distribution transformer's loading via scheduling microgrid resources in an efficient and cost-effective manner. Numerical simulations were carried out for various conditions of transformer loading to show the advantages and the effectiveness of the proposed model. A set of case studies on a testbed utility-owned microgrid demonstrate the effectiveness of the proposed model to reshape the loading of the distribution transformer at the point of interconnection in order to increase its RUL. The results showed that the utility companies can efficiently manage their resources to decrease transformer loss of useful life and consequently ensure a considerable increase in transformer's RUL.

References

1 X. Zhang and E. Gockenbach: "Asset-management of transformers based on condition monitoring and standard diagnosis," *IEEE Electr. Insul. Mag.*, vol. 24, no. 4, pp. 26–40, July/Aug. 2008.

2 A. E. B. Abu-Elanien and M. M. A. Salama, "Asset management techniques for transformers," *Electr. Power Syst. Res.*, vol. 80, no. 4, pp. 456–464, 2010.

3 H. Ma, T. K. Saha, C. Ekanayake, and D. Martin, "Smart transformer for smart grid intelligent framework and techniques for power transformer asset management," *IEEE Trans. Smart Grid*, vol. 6, no. 2, pp. 1026–1034, Mar. 2015.

4 A. Abiri-Jahromi, M. Parvania, F. Bouffard, and M. Fotuhi-Firuzabad, "A two-stage framework for power transformer asset maintenance management—part I: models and formulations," *IEEE Trans. Power Syst.*, 28, (2), pp. 1395–1403, 2013.

5 M. Arshad, S. M. Islam, and A. Khaliq, "Power transformer asset management," in *Proc. Int'l. Conf. Power System Technology (PowerCon)*, Singapore, pp. 1395–1398, Nov. 2004.

6 M. Žarković and Z. Stojković, "Analysis of artificial intelligence expert systems for power transformer condition monitoring and diagnostics," *Electr. Power Syst. Res.*, vol. 149, pp. 125–136, 2017.

7 K. T. Muthanna, A. Sarkar, K. Das, and K. Waldner, "Transformer insulation life assessment," *IEEE Trans. Power Delivery*, vol. 21, no. 1, pp. 150–156, Jan. 2006.

8 A. Hilshey, P. Hines, P. Rezaei, and J. Dowds, "Estimating the impact of electric vehicle smart charging on distribution transformer aging," *IEEE Trans. Smart Grid*, vol. 4, no. 2, pp. 905–913, June 2013.

9 V. Sarfi, S. Mohajeryami, and A. Majzoobi, "Estimation of water content in a power transformer using moisture dynamic measurement of its oil," *IET High Voltage*, vol. 2, no. 1, pp. 11–16, 2017.

10 J. Singh, Y. R. Sood, and P. Verma, "Experimental investigation using accelerated aging factors on dielectric properties of transformer insulating oil," *Electr. Power Compon. Syst.*, vol. 39, pp. 1045–1059, Jan. 2011.

11 A. Seier, P. D. H. Hines, and J. Frolik, "Data-driven thermal modeling of residential service transformers," *IEEE Trans. Smart Grid*, vol. 6, no. 2, pp. 1019–1025, Mar. 2015.

12 IEEE Std C57.91-2011, IEEE guide for loading mineral-oil-immersed transformers and step-voltage regulators, 2012.

13 M. Mahoor, A. Majzoobi, Z.S. Hosseini, and A. Khodaei, "Leveraging sensory data in estimating transformer lifetime," *North American Power Symposium (NAPS)*, Morgantown, WV, Sept. 2017.

14 M. Arshad and S. Islam, "A novel Fuzzy logic technique for power asset management," in *Proc. Industry Applications Conference*, Tampa, FL, pp. 276–286, Oct. 2006.

15 A. Majzoobi, M. Mahoor, and A. Khodaei, "Machine learning applications in estimating transformer loss of life," *IEEE PES General Meeting*, Chicago, IL, July 2017.

16 M. Mahoor and A. Khodaei, "Data fusion and machine learning integration for transformer loss of life estimation," in *IEEE PES Transmission and Distribution Conference and Exposition (T&D)*, Denver, CO, Apr. 2018.

17 J. L. Velasquez-Contreras, M. A. Sanz-Bobi, and S. Galceran Arellano, "General asset management model in the context of an electric utility: application to power transformers," *Electr. Power Syst. Res.*, vol. 81, no. 11, pp. 2015–2037, Nov. 2011.

18 V. Aravinthan and W. Jewell, W., "Controlled electric vehicle charging for mitigating impacts on distribution assets," *IEEE Trans. Smart Grid*, vol. 6, no. 2, pp. 999–1009, Mar. 2015.

19 C. Roe, F. Evangelos, J. Meisel, A.P. Meliopoulos, and T. Overbye, "Power system level impacts of PHEVs," in *Proc. 42nd Hawaii Int. Conf. Syst. Sci.*, Waikoloa, HI, pp. 1–10, 2009.

20 M. K. Gray and W. G. Morsi, "On the impact of single-phase plug-in electric vehicles charging and rooftop solar photovoltaic on distribution transformer aging," *Electr. Power Syst. Res.*, vol. 148, pp. 202–209, 2017.

21 M. K. Gray and W. G. Morsi, "On the role of prosumers owning rooftop solar photovoltaic in reducing the impact on transformer's aging due to plug-in electric vehicles charging," *Electr. Power Syst. Res.*, vol. 143, pp. 563–572, 2017.

22 S. Shokrzadeh, H. Ribberink, I. Rishmawi, and W. Entchev, "A simplified control algorithm for utilities to utilize plug-in electric vehicles to reduce distribution transformer overloading," *Energy*, vol. 133, pp. 1121–1131, 2017.

23 E. Ramos Munoz, G. Razeghi, L. Zhang, and F. Jabbari, "'Electric vehicle charging algorithms for coordination of the grid and distribution transformer levels," *Energy*, vol. 113, pp. 930–942, 2016.

24 Department of Energy Office of Electricity Delivery and Energy Reliability, "Smart Grid R&D Program-Summary Report: 2012 DOE Microgrid Workshop," https://www.energy.gov/sites/prod/files/2012%20Microgrid%20Workshop%20Report%2009102012.pdf. [Accessed 02-Apr-2018].

25 A. Majzoobi and A. Khodaei, "Application of microgrids in supporting grid flexibility," *IEEE Trans. Power Syst.*, vol. 32, no. 5, pp. 3660–3669, Sept. 2017.

26 L. Shi, Y. Luo, and G.Y. Tu, "Bidding strategy of microgrid with consideration of uncertainty for participating in power market," *Int. J. Electr. Power Energy Syst.*, vol. 59, pp. 1–13, 2014.

27 A. Majzoobi and S. Khodaei, "Application of microgrids in providing ancillary services to the utility grid," *Energy*, vol. 123, pp. 555–563, Mar. 2017.

28 A. Khodaei, S. Bahramirad, and M. Shahidehpour, "Microgrid planning under uncertainty," *IEEE Trans. Power Syst.*, vol. 30, no. 5, pp. 2417–2425, Sept. 2015.

29 A. Khodaei, "Microgrid optimal scheduling with multi-period islanding constraints," *IEEE Trans. Power Syst.*, vol. 29, no. 3), pp. 1383–1392, May 2014.

30 A. Khodaei, "Resiliency-oriented microgrid optimal scheduling," *IEEE Trans. Power Syst.*, vol. 5, no. 4, pp. 1584–1591, July 2014.

31 S. Bahramirad, W. Reder, and A. Khodaei, "Reliability-constrained optimal sizing of energy storage system in a microgrid," *IEEE Trans. Smart Grid*, Special Issue on Microgrids, vol. 3, no. 4, pp. 2056–2062, Dec. 2012.

32 M. Shahidehpour and J. Clair, "A functional microgrid for enhancing reliability, sustainability, and energy efficiency," *Electr. J.*, vol. 25, no. 8, pp. 21–28, Oct. 2012.

33 "DOE microgrid workshop report," www.energy.gov [Accessed 02-Apr-2018].

34 "Microgrid deployment tracker 4Q17," www.navigantresearch.com [Accessed 02-Apr-2018].

35 S. Parhizi, H. Lotfi, A. Khodaei, and S. Bahramirad, "State of the art in research on microgrids: a review," *IEEE Access*, vol. 3, pp. 890–925, July 2015.

36 A. M. Geoffrion, "Generalized benders decomposition," *J. Optim. Theory Appl.*, vol. 10, no. 4, pp. 237–260, 1972.

37 J. C. Smith and Z. C. Taskin, "A tutorial guide to mixed-integer programming models and solution techniques," *Optim. Med. Biol.*, pp. 521–548, 2008.

38 O. K. Gupta and A. Ravindran, "Branch and bound experiments in convex nonlinear integer programming," *Manag. Sci.*, vol. 31, no. 12, pp. 1533–1546, 1985.

39 M. Shahidehpour and Y. Fu, "Benders decomposition," *IEEE Power Energ. Mag.*, vol. 3, no. 2, pp. 20–21, Mar. 2005.

40 A. J. Conejo, E. Castillo, R. Minguez, and R. Garcia-Bertrand, "Decomposition Techniques in Mathematical Programming: Engineering and Science Applications," Springer Science & Business Media, 2006.

41 "PEAK Lab," https://portfolio.du.edu/peaklab/page/63877 [Accessed 02-Apr-2018].

42 "Weather history & data archive," https://www.wunderground.com/history/airport [Accessed 02-Apr-2018].

10

Dynamics of Microgrids in Distribution Network Flexibility

10.1   Principles of Distribution Network Flexibility

The growing trend of renewable generation installations driven primarily by current renewable portfolio standards, efficiency incentives, net metering, and the falling cost of renewable generation technologies [1, 2], challenges traditional practices in balancing electricity supply and demand and calls for innovative methods to reduce impacts on grid stability and reliability. Figure 10.1 shows daily net load (i.e. the consumer load minus local generation) variations in California ISO, the so-called duck curve, as an example of this challenge [3]. As renewable generation reached the 33% renewable target by 2020, the power grid required increased levels of fast ramping units to address abrupt changes (as much as 13 GW in three hours) in the net load, caused by concurrent fall in renewable generation and increase in demand.

To maintain system supply-demand balance, grid operators traditionally rely on bulk power generation resources, such as fast-ramping hydro and thermal units, that can be quickly dispatched and ramped up. These units, however, are limited in number and capacity, capital-intensive, time-consuming to construct, and subject to probable transmission network congestions. Addressing the variability of renewable generation, on the other hand, has long been an attractive area of research to complement renewable generation forecasting efforts [4]. Uncertainty considerations in power system operation and planning have significantly increased in the past few years as a large amount of uncertainty sources are integrated into power systems as a result of renewable generation proliferation. The renewable generation integration problem can be investigated under two contexts: large-scale (which attempts to manage the generation of wind and solar farms) [5–8] and small-scale (which deals with renewable generation at the distribution level). Small-scale coordination approaches mainly focus on various methods of demand-side management, such as demand response [9–12], energy storage

The Economics of Microgrids, First Edition. Amin Khodaei and Ali Arabnya.
© 2024 The Institute of Electrical and Electronics Engineers, Inc.
Published 2024 by John Wiley & Sons, Inc.

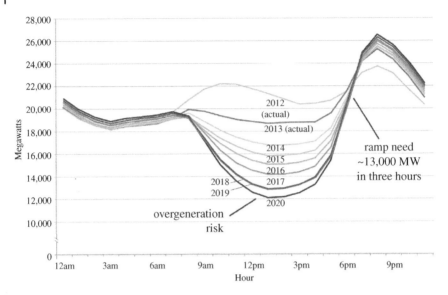

Figure 10.1 The current and future estimates of over-generation and ramping effect in California [3].

[13–17], and aggregated electric vehicles [18, 19]. However, these methods each encounter obstacles that may prevent a viable application, such as the need for advanced metering infrastructure in deploying demand response, privacy and customer willingness issues in electric vehicle applications, and financial aspects in energy storage deployment. Leveraging available flexibility in existing microgrids for addressing renewable generation integration, as proposed in [20, 21] and extended in this chapter, will offer a potentially more viable solution to be used in distribution networks, and thus calls for additional studies. Microgrids have been significantly deployed over the past few years and are anticipated to grow even more in the near future [22–30], at both national and international levels [31], where future power grids can be pictured as systems of interconnected microgrids [32].

In this chapter, the flexibility advantage of microgrids as a complementary value proposition in grid support is presented. The microgrid's capability in managing its power exchange with the utility grid in the grid-connected mode for mitigating the net load ramping in the distribution network is elaborated. There are several studies that investigate how a microgrid can participate in the upstream network market and offer services to the grid. In [33], an optimal bidding strategy via a microgrid aggregator is proposed to involve all small-scale microgrids in an electricity market via real-time balancing market bidding. In [34], an optimal bidding strategy based on two-stage stochastic linear programming for an electric vehicle aggregator who participates in the day-ahead energy and regulation markets is proposed. Further, the market conditions and the associated uncertainty of the electric vehicle fleet are incorporated in their study. Reference [35] proposes a two-stage market model for microgrid power exchange with the utility grid via an aggregator to achieve an efficient market equilibrium. Reference [36] presents a risk-constrained optimal hourly bidding model for microgrid aggregators to incorporate various uncertainties and maximize the microgrid benefit. The study in [37] proposes an optimal dispatch strategy for the residential loads via artificial neural network for calculating the demand forecast error when the demand changes are known one hour ahead with respect to the day-ahead forecasted values. Reference [38] presents a stochastic bidding strategy for microgrids participating in energy and spinning reserve markets, considering the load and renewable generation uncertainty. In [39], a stochastic look-ahead economic dispatch model for near-real-time power system operations is proposed and its benefits and implementation feasibility for assessing the power system's economic risk are further explored. These research works primarily rely on a market mechanism to procure microgrids' flexibility and accordingly capture the unbalanced power in the day-ahead market as well as the ramping and variabilities caused by forecast errors or unforeseen real-time events. In this chapter, however, this problem is studied from a microgrid perspective, i.e. how a microgrid controller can manage local resources to offer required/desired services to the utility grid. This work is particularly important in networks where a market mechanism cannot be established but grid operators are interested in low-cost and distributed solutions for managing grid flexibility. The main contributions of this chapter are listed as follows:

- A flexibility-oriented microgrid optimal scheduling model is developed to optimally manage local microgrid resources while providing flexibility services to the utility grid. This model is achieved by transforming the distribution net load variability limits into constraints on the microgrid net load.
- A coordinated grid-connected and islanded operation is considered in the model development to take into account microgrid's potential islanding while supporting the utility grid in the grid-connected mode.
- A high-resolution operation is modeled via consideration of both intra-hour and inter-hour time periods, which is capable of integrating quick variations in renewable generation.

Unlike existing studies on distribution network flexibility procurement, which focus on microgrid participation in grid support via a market mechanism, we investigate the problem from a microgrid's perspective, i.e. how a microgrid controller can manage local resources to offer required/desired services to the utility grid.

10.2 Economic Variables for Microgrids in Electricity Distribution Networks

Consider a distribution feeder consisting of a set of $N = \{1, 2, ..., N\}$ customers (both consumers and prosumers) and one microgrid. The net load of each customer $j \in$ N and the microgrid are denoted by P^c_{jtk} and P^M_{tk}, respectively, where t is the inter-hour time index and k is the intra-hour time index, as demonstrated in Figure 10.2. To fully supply the total net load in this feeder, a power of P^u_{tk} needs to be provided by the utility grid where:

$$P^u_{tks} = P^M_{tks} + \sum_{j \in N} P^c_{jtks} \qquad \forall t, \forall k, \forall s, \tag{10.1}$$

where c is superscript for distribution network consumers and prosumers, u is superscript for the utility grid, and s is index for scenarios. To address the net load variability seen by the grid operator, the intra-hour variability (10.2) and inter-hour variability (10.3) in the utility grid power will need to be constrained, as follows:

$$\left| P^u_{tks} - P^u_{t(k-1)s} \right| \leq \Delta_1 \qquad \forall t, \forall s, k \neq 1, \tag{10.2}$$

$$\left| P^u_{t1s} - P^u_{(t-1)Ks} \right| \leq \Delta_2 \qquad \forall t, \forall s, \tag{10.3}$$

where Δ_1 is intra-hour flexibility limit, and Δ_2 is inter-hour flexibility limit. These limits are selected by the grid operator based on the day-ahead net load forecasts and desired grid flexibility during each time interval. There are various methods to determine the grid flexibility [6, 40–42]. If this calculated flexibility is less than the required grid flexibility, which is obtained based on net load forecasts, the grid operator can utilize distributed resources, such as microgrids, to compensate for the shortage in grid flexibility. Therefore, intra- and inter-hour limits will be obtained by comparing the available and required grid flexibility. Considering the importance of grid flexibility limits on the microgrid operation, a system-level study needs to be performed by the utility company. This topic will be investigated in a follow-up research. The grid operator can furthermore calculate these limits

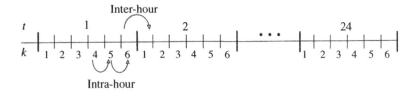

Figure 10.2 The schematic diagram of inter-hour and intra-hour time intervals.

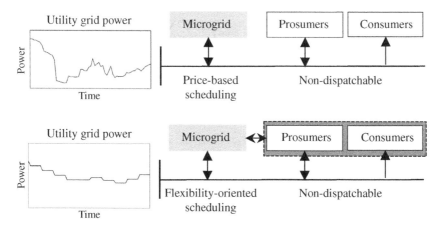

Figure 10.3 Impact of the microgrid in increasing the net load variabilities (top) or capturing the variabilities (bottom).

using a cost–benefit analysis, i.e. to upgrade the current infrastructure to address increasing flexibility requirements or to procure the flexibility of existing microgrids and in turn pay for their service. This topic, however, requires further analysis and modeling, which will be carried out in follow-up research.

Figure 10.3 shows the schematic diagram of a feeder consisting of a microgrid along with other connected loads. The microgrid can be scheduled based on price considerations, i.e. local resources are scheduled in a way that the microgrid operation cost is minimized during the grid-connected mode (Figure 10.3, top). The only factor impacting the microgrid scheduling results from the utility grid side is the real-time electricity price (hence the term price-based scheduling). Price-based scheduling can potentially exacerbate the consumption variability. On the other hand, microgrid resources can be scheduled in coordination with other loads in the same distribution feeder, and thus support the utility grid in mitigating potential variabilities and ensuring supply–load balance (Figure 10.3, bottom). Although the objective is still to minimize the operation cost during the grid-connected mode, this scheduling is primarily based on the grid flexibility requirements (hence the term flexibility-oriented scheduling).

10.3 Economic Models for Distribution Network Operations Under Microgrid Dynamics

The microgrid optimal scheduling problem aims at determining the least-cost schedule of available resources (distributed energy resources [DERs] and loads) while considering prevailing operational constraints, i.e.

$$\min \sum_t \sum_k \tau \left[\sum_{i \in G} F_i(P_{itk0}) + \rho_t^M P_{tk0}^M \right] + \sum_t \sum_k \sum_s \tau \, \psi_s \, \lambda \, LS_{tks} \qquad (10.4)$$

s.t.

$$\sum_i P_{itks} + P_{tks}^M + LS_{tks} = \sum_d D_{dtks} \qquad \forall t, \forall k, \forall s, \qquad (10.5)$$

$$-P^{M,\max} w_{tks} \leq P_{tks}^M \leq P^{M,\max} w_{tks} \qquad \forall t, \forall k, \forall s, \qquad (10.6)$$

$$\{P_{itks}, D_{dtks}\} \in O_s \qquad \forall i, \forall t, \forall k, \forall s, \qquad (10.7)$$

$$P_{tks}^M \in F_s \qquad \forall t, \forall k, \forall s, \qquad (10.8)$$

where d is index for loads, i is index for DERs, k is index for intra-hour time periods, s is index for scenarios, t is index for inter-hour time periods, F is a set of flexibility constraints, G is a set of dispatchable units, O is a set of operation constraints, P is DER output power variable, P^M is variable for utility grid power exchange with the microgrid, LS is load curtailment variable, D is variable for load demand, w is a binary islanding indicator parameter (1 if grid-connected, 0 if islanded), $F(.)$ is generation cost, τ is parameter for time period, ψ is probability of islanding scenarios, and λ is value of lost load (VOLL). Objective (10.4) minimizes the microgrid's daily operations cost, which includes the local generation cost, cost of energy transfer with the utility grid, and the outage cost. The outage cost (also known as the cost of unserved energy) is defined as the load curtailment times the VOLL. The VOLL represents the customers' willingness to pay for reliable electricity service and to avoid power outages, which can also be perceived as the energy price for compensating curtailed loads. The VOLL depends on the type of customers, time and duration of outage, time of advanced notification of outage, and other specific traits of an outage [43]. The load balance Eq. (10.5) ensures that the sum of the injected/withdrawn power from the utility grid and local DERs (i.e. dispatchable units, non-dispatchable units, and distributed energy storage) would match the microgrid load. The load curtailment variable is used to ensure a feasible solution in the islanded operation if adequate generation is not available. The power of energy storage can be negative (charging), positive (discharging), or zero (idle). Since the power can be exchanged between the utility grid and the microgrid, P_{tks}^M can be positive (power import), negative (power export), or zero. The power transfer with the utility grid is limited by (10.6). The binary islanding parameter (which is 1 when grid-connected and 0 when islanded) ensures that the microgrid interacts with the utility grid only during the grid-connected operation. Microgrid DERs, loads, and the main grid power transfer are further subject to operation and flexibility constraints, respectively, represented by sets O_s and F_s in (10.7) and (10.8).

10.3.1 Operation Constraints (O_s)

The microgrid components to be modeled in the optimal scheduling problem include DERs (i.e. generation units and energy storage) and loads. Microgrid loads are categorized into two types: fixed (which cannot be altered and must be satisfied under normal operation conditions) and adjustable (which are responsive to price variations and/or control signals). Generation units in a microgrid are either dispatchable (i.e. units that can be controlled by the microgrid controller) or non-dispatchable (i.e. wind and solar units that cannot be controlled by the microgrid controller since the input source is uncontrollable). The primary applications of the energy storage are to coordinate with generation units for guaranteeing microgrid generation adequacy, energy shifting, and islanding support. From these microgrid components, only dispatchable DGs, energy storage, and adjustable loads can provide flexibility benefits for the microgrid due to their controllability. Microgrid component constraints are formulated as follows:

$$P_i^{\min} I_{it} \leq P_{itks} \leq P_i^{\max} I_{it} \qquad \forall i \in G, \forall t, \forall k, \forall s, \tag{10.9}$$

$$P_{itks} - P_{it(k-1)s} \leq UR_i \qquad \forall i \in G, \forall t, \forall s, k \neq 1, \tag{10.10}$$

$$P_{it1s} - P_{i(t-1)Ks} \leq UR_i \qquad \forall i \in G, \forall t, \forall k, \forall s, \tag{10.11}$$

$$P_{it(k-1)s} - P_{itks} \leq DR_i \qquad \forall i \in G, \forall t, \forall s, k \neq 1, \tag{10.12}$$

$$P_{i(t-1)Ks} - P_{it1s} \leq DR_i \qquad \forall i \in G, \forall t, \forall k, \forall s, \tag{10.13}$$

$$T_i^{\text{on}} \geq UT_i\left(I_{it} - I_{i(t-1)}\right) \qquad \forall i \in G, \forall t, \tag{10.14}$$

$$T_i^{\text{off}} \geq DT_i\left(I_{i(t-1)} - I_{it}\right) \qquad \forall i \in G, \forall t, \tag{10.15}$$

$$P_{itks} \leq P_{itk}^{\text{dch, max}} u_{it} - P_{itk}^{\text{ch, min}} v_{it} \qquad \forall i \in S, \forall t, \forall k, \forall s, \tag{10.16}$$

$$P_{itks} \geq P_{itk}^{\text{dch, min}} u_{it} - P_{itk}^{\text{ch, max}} v_{it} \qquad \forall i \in S, \forall t, \forall k, \forall s, \tag{10.17}$$

$$u_{it} + v_{it} \leq 1 \qquad \forall i \in S, \forall t, \tag{10.18}$$

$$C_{itks} = C_{it(k-1)s} - (P_{itks} u_{it} \tau / \eta_i) - P_{itks} v_{it} \tau \qquad \forall i \in S, \forall t, \forall s, k \neq 1, \tag{10.19}$$

$$C_{it1s} = C_{i(t-1)Ks} - (P_{it1s} u_{it} \tau / \eta_i) - P_{it1s} v_{it} \tau \qquad \forall i \in S, \forall t, \forall s, \tag{10.20}$$

$$C_i^{\min} \leq C_{itks} \leq C_i^{\max} \qquad \forall i \in S, \forall t, \forall k, \forall s, \tag{10.21}$$

$$T_{it}^{\text{ch}} \geq MC_i\left(u_{it} - u_{i(t-1)}\right) \qquad \forall i \in S, \forall t, \tag{10.22}$$

$$T_{it}^{\text{dch}} \geq MD_i\left(v_{it} - v_{i(t-1)}\right) \qquad \forall i \in S, \forall t, \tag{10.23}$$

$$D_d^{\min} z_{dtk} \leq D_{dtks} \leq D_d^{\max} z_{dtk} \qquad \forall d \in D, \forall t, \forall k, \forall s, \tag{10.24}$$

$$T_d^{\text{on}} \geq MU_d \left(z_{dt} - z_{d(t-1)} \right) \qquad \forall d \in D, \forall t, \tag{10.25}$$

$$\sum_{[\alpha, \beta]} D_{dtks} = E_d \qquad \forall d \in D, \forall s, \tag{10.26}$$

where ch is superscript for energy storage charging, dch is superscript for energy storage discharging, D is a set of adjustable loads, S is a set of energy storage systems, I represents a variable for commitment state of dispatchable units, C is a variable for energy storage available (stored) energy, T^{on} is a variable for number of successive ON hours, T^{off} is a variable for number of successive OFF hours, T^{ch} is a variable for number of successive charging hours, T^{dch} is a variable for number of successive discharging hours, u is a variable for energy storage discharging state (1 when discharging, 0 otherwise), v is a variable for energy storage charging state (1 when charging, 0 otherwise), z is a variable for adjustable load state (1 when operating, 0 otherwise), E is a parameter for load's total required energy, MC is a parameter for minimum charging time, MD is a parameter for minimum discharging time, UR is a parameter for ramp-up rate parameter, DR is a parameter for ramp-down rate parameter, UT is a parameter for minimum uptime, DT is a parameter for minimum downtime, MU is a parameter for minimum operating time, α and β are specified start and end times of adjustable loads, ρ is market price, and η is energy storage efficiency. Constraint (10.9) represents the maximum and minimum generation capacities of dispatchable units. Dispatchable generation units are also subject to ramp-up and ramp-down constraints, which are defined by (10.10)–(10.13). Equations (10.10) and (10.12) represent the ramping constraints for intra-hour intervals, while (10.11) and (10.13) represent the ramping constraints for inter-hour intervals. The minimum up and downtime limits are imposed by (10.14) and (10.15), respectively. The minimum and maximum limits of the energy storage charging and discharging, based on the operation mode, are defined by (10.16) and (10.17), respectively. While charging, the binary charging state v is one and the binary discharging state u is zero; while in the discharging mode, the binary charging state v is zero and the binary discharging state u is one. The energy storage charging power is a negative value, which is compatible with the negative amount for limitations of constraints (10.16) and (10.17) for the charging mode. Only one of the charging or discharging modes at every time period is possible, which is ensured by (10.18). The stored energy is calculated based on the available stored energy and the amount of charged/discharged power, which are represented in (10.19) and (10.20) for intra-hour and inter-hour intervals, respectively. The time period of charging and discharging is considered to be $\tau = (1/K)h$, where K is the number of intra-hour periods and h represents a time period of one hour. The amount of stored energy in energy storage is restricted with its capacity (10.21). The minimum charging and discharging times are represented in (10.22) and (10.23), respectively. Adjustable loads are subject to minimum and maximum rated powers (10.24), where binary operating state z is 1 when load is consuming power and 0 otherwise. The minimum operating time (10.25) and the required energy to complete an operating cycle (10.26) are further considered for adjustable loads. It is worth mentioning that $t = 0$, which would appear in (10.3), (10.14), (10.15), (10.22), and (10.23), represents the last hour of the previous scheduling horizon, here $t = 24$.

10.3.2 Flexibility Constraints (F$_s$)

Flexibility constraints represent additional limits on the microgrid's power exchange with the utility grid. These constraints are defined in such a way that the microgrid net load is matched with the aggregated net load of connected prosumers/consumers to offset likely variations. To obtain the flexibility constraints, the value of P^u_{tks}, i.e. $P^u_{tks} = P^M_{tks} + \sum_{j \in N} P^c_{jtks}$, is substituted in (10.2) and (10.3). By proper rearrangements, the inter-hour and intra-hour flexibility constraints will be accordingly obtained as in (10.27) and (10.28):

$$
-\Delta_1 - \left(\sum_j P^c_{jtk} - \sum_j P^c_{jt(k-1)} \right) \leq P^M_{tks} - P^M_{t(k-1)s} \leq \Delta_1
$$
$$
- \left(\sum_j P^c_{jtk} - \sum_j P^c_{jt(k-1)} \right) \qquad \forall t, \forall s, k \neq 1,
$$
(10.27)

$$
-\Delta_2 - \left(\sum_j P^c_{jt1} - \sum_j P^c_{j(t-1)K} \right) \leq P^M_{t1s} - P^M_{(t-1)Ks} \leq \Delta_2
$$
$$
- \left(\sum_j P^c_{jt1} - \sum_j P^c_{j(t-1)K} \right) \qquad \forall t, \forall s.
$$
(10.28)

Accordingly, new time-dependent flexibility limits can be defined as follows:

$$
\Delta^{low}_{1,tk} = -\Delta_1 - \left(\sum_j P^c_{jtk} - \sum_j P^c_{jt(k-1)} \right) \qquad \forall t, k \neq 1,
$$
(10.29)

$$
\Delta^{up}_{1,tk} = \Delta_1 - \left(\sum_j P^c_{jtk} - \sum_j P^c_{jt(k-1)} \right) \qquad \forall t, k \neq 1,
$$
(10.30)

$$\Delta_{2,t}^{low} = -\Delta_2 - \left(\sum_j P_{jt1}^c - \sum_j P_{j(t-1)K}^c \right) \qquad \forall t, \tag{10.31}$$

$$\Delta_{2,t}^{up} = \Delta_2 - \left(\sum_j P_{jt1}^c - \sum_j P_{j(t-1)K}^c \right) \qquad \forall t, \tag{10.32}$$

where Δ_1^{low} and Δ_1^{up} are microgrid time-dependent intra-hour lower and upper flexibility limits, respectively, and Δ_2^{low} and Δ_2^{up} are microgrid time-dependent inter-hour lower and upper flexibility limits, respectively. These new constraints convert the required flexibility by the grid operator into a limit on the microgrid net load. Although utility grid flexibility limits, i.e. Δ_1 and Δ_2, are constant and determined by the grid operator, the limits on the microgrid net load are highly variable as they comprise the aggregated net load of all N customers in the distribution feeder. Depending on the considered time resolution for forecasts, these limits can change from every one minute to every one hour in the scheduling horizon. The flexibility limits can be adjusted by the grid operator to achieve the desired net load in the distribution network. For example, a value of zero for Δ_1 would eliminate intra-hour variations.

It is worth mentioning that connected prosumers/consumers are considered as given parameters (forecasted) in the optimization problem. There will be no direct communication between the microgrid and the connected prosumers/consumers. Rather, all communications will be managed through the grid operator. Therefore, the microgrid only communicates with the grid operator and sends/receives the required data for capturing and mitigating the distribution network's net load variabilities.

10.3.3 Islanding Considerations

The islanding is performed to rapidly disconnect the microgrid from a faulty distribution network, safeguard the microgrid components from upstream disturbances, and protect voltage-sensitive loads when a quick solution to utility grid voltage problems is not imminent. The time and duration of such disturbances, however, are not known to microgrids in advance. Islanding is considered via a Θ-k islanding criterion, where Θ (i.e. equal to $T \times K$) represents the total number of intra-hour time periods in the scheduling horizon and k represents the number of consecutive intra-hour periods that the microgrid should operate in the islanded mode. To apply this criterion to the proposed model, the binary islanding indicator w is defined and added to the microgrid power exchange constraint (10.6). Several scenarios are defined based on the number of intra-hour time periods (for instance, 144 scenarios for 10-minute intra-hour periods), and the value of w in each scenario is obtained based on the Θ-k islanding criterion, i.e. in each scenario, w will

t	1						2						3						
k	1	2	3	4	5	6	1	2	3	4	5	6	1	2	3	4	5	6	...
Scenario 1	0	0	0	0	1	1	1	1	1	1	1	1	1	1	1	1	1	1	...
Scenario 2	1	0	0	0	0	1	1	1	1	1	1	1	1	1	1	1	1	1	...
Scenario 3	1	1	0	0	0	0	1	1	1	1	1	1	1	1	1	1	1	1	...
Scenario 4	1	1	1	0	0	0	0	1	1	1	1	1	1	1	1	1	1	1	...
Scenario 5	1	1	1	1	0	0	0	0	1	1	1	1	1	1	1	1	1	1	...

Figure 10.4 First five islanding scenarios associated with a Θ-4 islanding criterion.

be 0 for k consecutive intra-hour time periods (imposing an islanded operation) and 1 in other periods (representing the grid-connected operation). Figure 10.4 shows the first five islanding scenarios, from a total of 144 scenarios, associated with a Θ-4 islanding criterion, which requires that the microgrid be able to operate in the islanded mode for any four consecutive intra-hour periods once it is switched to the islanded mode. Further discussions on the Θ-k islanding criterion can be found in [23]. It should be noted that the proposed model is generic and can be applied to any microgrid size without loss of generality.

10.4 Case Study

A microgrid test system with four dispatchable units, two non-dispatchable units including wind and solar, one energy storage, and five adjustable loads is used to study the performance of the proposed model. The characteristics of the microgrid DERs and loads and the hourly market price are borrowed from [23]. The maximum ramping capability of the microgrid, based on the maximum ramping capacity of DERs, is 18 MW/h, and the capacity of the line connecting the microgrid to the distribution feeder is assumed to be 10 MW. A VOLL of $10,000/MWh is considered for the microgrid.

The aggregated consumption profile of consumers/prosumers connected to the system in the same feeder as the microgrid is shown in Figure 10.5. As shown, it consists of aggregated values for the distributed solar generation, consumption, and the net load (i.e. difference between the local consumption and generation). The net load should be supplied by the utility grid, and as the figure demonstrates, it includes considerable variability due to the local solar generation. The maximum ramping of this net load is 3.3 MW/10-min, and the peak net load is 12.9 MW. This net load variability should be satisfied by either fast response units deployed by the utility or locally by the microgrid, where the latter is discussed here. The proposed flexibility-oriented microgrid optimal scheduling model is developed using mixed-integer programming and solved using CPLEX 12.6. It should be noted that the

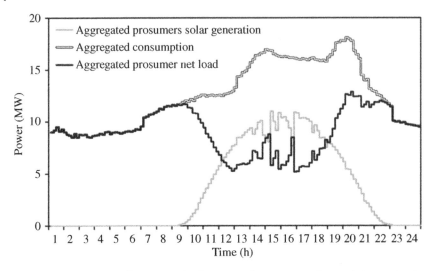

Figure 10.5 Aggregated prosumers' solar generation, consumption, and the net load in the distribution feeder.

computation time for the studied cases was between three and four minutes, with an average of 3 minutes and 22 seconds.

Case 1: The grid-connected, price-based optimal scheduling is analyzed for a 24-hour planning horizon. The price-based scheduling denotes that the microgrid seeks to minimize its operations costs and does not have any commitment to support the utility grid in capturing distribution network net load variabilities. Table 10.1 shows the schedule of dispatchable units and the energy storage for 24 hours of operation in this case. A commitment state of 1 represents that the dispatchable unit is ON, while 0 represents that the unit is not committed. The energy storage charging, discharging, and idle states are represented by −1, 1, and 0, respectively. The bold values in Table 10.1 represent changes in the schedule due to the islanding requirements. Dispatchable unit 1 has the lowest operation cost, so it is committed in all scheduling hours, while other units are committed and dispatched when required based on economic and reliability considerations. It should be noted that the amount of load curtailment during the islanded operation is considered as a measure of microgrid reliability. The energy storage is charged in low-price hours and discharged in high-price hours, i.e. an energy arbitrage, to maximize the benefits and minimize the operation cost. As Table 10.1 shows, the islanding criterion leads to the commitment of more units in the grid-connected mode to guarantee a seamless islanding.

Figure 10.6 depicts the microgrid net load and the distribution feeder net load (i.e. the microgrid net load plus the aggregated consumer/prosumer net load in

Table 10.1 DER schedule in case 1.

											Hours (1–24)												
G1	1	1	1	1	1	1	1	1	1	1	1	1	1	1	1	1	1	1	1	1	1	1	1
G2	1	1	1	1	1	1	1	1	1	1	1	1	1	1	1	1	1	1	1	1	1	1	1
G3	1	1	1	0	0	0	0	1	1	1	1	1	1	1	1	1	1	1	1	1	1	1	0
G4	0	0	0	0	0	0	0	0	0	0	0	1	1	1	1	0	1	1	1	1	1	1	0
DES	−1	−1	−1	−1	0	0	0	0	0	0	0	0	0	0	0	0	0	0	0	0	1	0	0

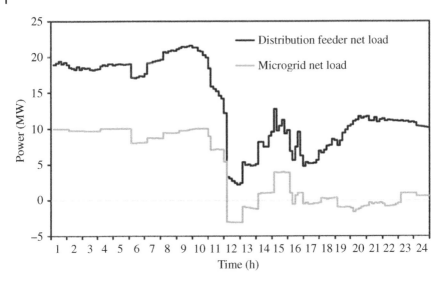

Figure 10.6 Distribution feeder net load, and microgrid net load for the 24-hour horizon in Case 1.

Figure 10.5). As shown, the microgrid imports the power from the utility grid in low-price hours and switches over to local generation when the utility grid price is high. This scheduling causes a 21.58 MW peak load for the utility grid between hours 9 and 10 (that is a new morning peak) and exacerbates the distribution feeder ramping requirement (which is increased to 8.9 MW/10-min between hours 11 and 12 in this case). In addition, the net load variability is significantly increased in this case. Therefore, the utility grid encounters a sharp net load ramping and variations, which are primarily caused by the microgrid. This result indicates that the microgrid can potentially have a negative impact on the distribution network's net load when scheduled only based on the price data and economic considerations. The microgrid operations cost in this case is $11,748.3.

Case 2: In this case, the flexibility-oriented microgrid optimal scheduling is made, rather than the price-based scheduling, to support the utility grid in addressing net load variations. A Θ-1 islanding criterion with 10-min intra-hour periods is considered. This islanding criterion ensures that the microgrid is capable of switching to the islanded mode to reliably supply local loads (for any 10-min islanding during the scheduling horizon), while supporting the utility grid by providing required flexibility during the grid-connected operation. The flexibility limits of 0.5 MW/10-min are considered for inter-hour and intra-hour ramping. The intra-hour and inter-hour ramping constraints are accordingly developed, as modeled in (10.27)–(10.32). Table 10.2 shows the derived schedules of dispatchable

Table 10.2 DER schedule in case 2.

	Hours (1–24)																							
G1	1	1	1	1	1	1	1	1	1	1	1	1	1	1	1	1	1	1	1	1	1	1	1	1
G2	1	1	1	1	1	1	1	1	1	1	1	1	1	1	1	1	1	1	1	1	1	1	1	1
G3	1	1	1	0	0	0	1	1	1	1	1	1	1	1	1	1	1	1	1	1	1	1	1	1
G4	0	0	0	0	0	0	0	0	0	0	1	**1**	**0**	0	0	0	1	1	1	1	1	1	0	0
DES	0	−1	−1	−1	−1	**−1**	0	0	0	0	1	1	0	1	1	**1**	**1**	**1**	1	1	1	1	0	0

units and the energy storage for the scheduling horizon. The bold figures in Table 10.2 represent changes in the schedule, while the highlighted cells represent changes in the dispatched power compared to Case 1. As shown, the commitment of unit 4 and the energy storage, as well as the dispatched power of all DERs, are changed compared to Case 1 to satisfy the flexibility constraints. These changes in the schedules increase the microgrid operations cost to $12,077. The difference between this cost and the microgrid operations cost in Case 1 should be paid to the microgrid, as a minimum, to incentivize the microgrid for providing flexibility and supporting the utility grid.

Figure 10.7 shows the distribution feeder net load and the microgrid net load in this case. Comparison of Figures 10.6 and 10.7 shows the positive impact of the microgrid in changing the distribution network net load in a way that is desirable for the utility grid. As Figure 10.6 illustrates, the distribution feeder net load, which should be supplied by the utility grid, consists of several ramping events in the order of a few MW/10-min as well as a severe ramping of 8.9 MW/10-min between hours 11 and 12. In Figure 10.7, however, all these variabilities are reduced to 0.5 MW/10-min as targeted by the grid operator. Moreover, Figure 10.8 depicts the ramping of the utility grid in both studied cases. This figure clearly demonstrates the effectiveness of the proposed model in reducing the distribution network net load ramping, as the obtained data from Case 2 is efficiently confined between the desired ramping values.

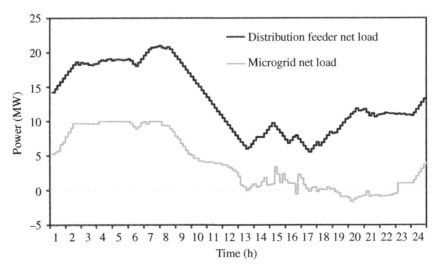

Figure 10.7 Distribution feeder net load and microgrid net load for 0.5 MW/10-min inter-hour and intra-hour utility ramping in Case 2.

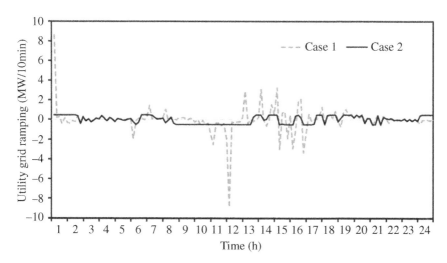

Figure 10.8 Utility grid net load ramping in the two studied cases.

The results in Case 2 show that to obtain the desired ramping, the microgrid needs to deviate from its price-based schedule. This deviation results in a $328.7 increase in the microgrid operations cost (i.e. $12,077–$11,748.3). This increase represents the microgrid's lost revenue. To incentivize the microgrid to opt in for offering flexibility services to the utility grid, the amount of incentive that should be paid to the microgrid must be equal to or greater than this amount. If less, the microgrid would economically prefer to find its price-based schedule while disregarding the grid requirements. However, it would be extremely beneficial for the utility grid to incentivize the microgrid; otherwise, the microgrid may exacerbate the distribution network net load variability as discussed in Case 1. It is worth mentioning that the microgrid's lost revenue is a function of the consumers/prosumers net load variations as well as values of Δ_1 and Δ_2, which are further investigated in the following.

Case 3: After demonstrating the effectiveness of the proposed model by comparing Cases 1 and 2, the impact of ramping limits is studied in this case. To illustrate that the microgrid is also capable of meeting tight ramping limits, a value of zero is considered for the intra-hour ramping and 2 MW/10-min for the inter-hour ramping. Figure 10.9 depicts the solution of this case. Considering a value of zero for intra-hour ramping completely eliminates the intra-hour variabilities in the distribution network net load; hence, the obtained consumption is constant within each operation hour while it can change by up to 2 MW between any two consecutive operation hours.

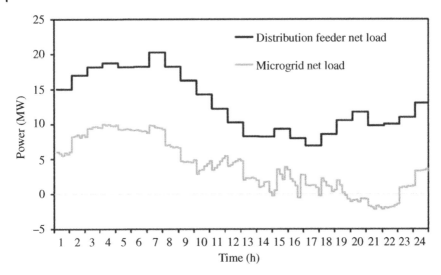

Figure 10.9 Distribution feeder net load and microgrid net load for 2 MW/10-min inter-hour and 0 MW/10-min intra-hour utility ramping.

To not violate the limits, the microgrid's imported power from the utility grid is decreased when the net load is increasing. Furthermore, microgrid's export to the utility grid in high-price hours is changed to support the ramping limits. For instance, the microgrid power export to the utility grid in hours 12–14, which was based on economic considerations, is now changed to power import from the utility grid. Figure 10.10 shows the obtained results of Figure 10.9 between hours 12 and 20, which better demonstrates the viable application of the microgrid in reducing the net load variability and sharp ramping.

The results from flexibility-oriented microgrid optimal scheduling for different amounts of inter-hour (changing between 0.5 and 5) and intra-hour (changing between 0 and 2) ramping limits are presented in Table 10.3. It should be noted that the derived solution is near-optimal, mainly due to nonlinearity of the original problem and presence of uncertainties.

The obtained results show that the microgrid operations cost is increased by reducing the inter-hour and intra-hour ramping limits; however, these changes are not linear. For example, the microgrid operations cost when the intra-hour ramping limit is 0 is considerably higher than in other cases. This is due to two main reasons, which are as follows: (i) the need to commit more units and dispatch them at uneconomical operation points, in a way that they can provide the required flexibility; and (ii) the possibility of load curtailment in the microgrid. The ramping limits are added as constraints to the problem, while the load

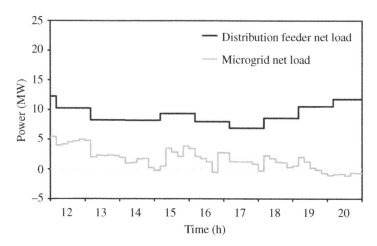

Figure 10.10 Distribution feeder net load and microgrid net load for 2 MW/10-min inter-hour and 0 MW/10-min intra-hour utility ramping during net load peak hours.

Table 10.3 Microgrid operation cost ($) for various ramping limits.

Inter-hour ramping limit Δ_2 (MW/10 min)	Intra-hour ramping limit Δ_1 (MW/10 min)			
	0	0.5	1	2
0.5	36,305.6	12,077	11,886.9	11,825.3
1	19,799.1	12,011.5	11,860.1	11,804
1.5	14,930.4	11,977.5	11,845.2	11,796.5
2	13,329.1	11,951.4	11,834.1	11,790.1
2.5	12,790.2	11,936.4	11,826.4	11,786
3	12,607.5	11,925.8	11,819.7	11,782.1
3.5	12,532.5	11,916.8	11,813.7	11,778.6
4	12,485.9	11,906.8	11,808.6	11,775.6
4.5	12,460.1	11,898.3	11,804.3	11,772.7
5	12,445.6	11,891	11,800.1	11,770.1

curtailment is added as a penalty to the objective function. It results in prioritizing the flexibility limit (i.e. problem feasibility) on the load curtailment (i.e. problem optimality). Additional measures are necessary to be considered in order to prevent load curtailment in the microgrid. The utility grid incentive in each case must at least cover the microgrid's lost revenue. According to Table 10.3, if the utility grid decides to eliminate the intra-hour ramping, it should pay at least \$24,557.3 and \$697.3 to the microgrid for Δ_2 values equal to 0.5 MW/10-min and 5 MW/10-min, respectively. Whereas, in the case of 2 MW/10-min as desired intra-hour ramping, at least \$77 and \$21.8 should be paid to the microgrid for Δ_2 equal to 0.5 MW/10-min and 5 MW/10-min, respectively. These results advocate for the importance of a cost–benefit analysis by the grid operator to determine the most suitable inter-hour and intra-hour ramping limits.

Microgrids can potentially be utilized in distribution networks as a solution for mitigating net load ramping and variability. Specific features of the proposed microgrid optimal scheduling model with multi-period islanding and flexibility constraints are as follows:

- *Flexibility consideration:* The inter-hour and intra-hour ramping constraints have been considered in the proposed model to ensure that the utility grid's desired power is obtained for different time resolutions.
- *Economical and reliable operations:* The proposed model determines the least-cost schedule of microgrid loads and DERs while supporting the utility grid in addressing net load ramping. In addition, the consideration of Θ-k islanding criterion ensures the microgrid's reliability in supplying local loads during the islanded mode.
- *High-resolution scheduling:* 10-minute time interval scheduling was considered in studied cases, which offers a high-resolution scheduling and is efficient for capturing net load variabilities. The proposed model offers the capability to consider various intra-hour time resolutions.
- *Localized and low-cost solution:* Using microgrids as local solutions for addressing distribution net load ramping can significantly reduce the utility grid's investments in upgrading the generation, transmission, and distribution facilities. This significant cost savings can be realized at the small expense of incentivizing microgrids to offer flexibility services.

10.5 Summary

Distributed renewable energy resources have attracted significant attention in recent years due to the falling cost of renewable energy technology, extensive government incentives, and new applications for improving load-point reliability. This growing proliferation, however, is changing the traditional consumption load curves by adding considerable levels of variability and further challenging the electricity supply–demand balance. In this chapter, we investigated an application of microgrids in effectively capturing the distribution network's net load variability, caused primarily by the prosumers. Microgrids provide a viable and localized solution to this challenge while removing the need for costly investments by the electric utility in reinforcing the existing electricity infrastructure. A flexibility-oriented microgrid optimal scheduling model was proposed to coordinate the microgrid net load with the aggregated consumers/prosumers net load in the distribution network with a focus on ramping issues. The proposed coordination is performed to capture both inter-hour and intra-hour net load variabilities. The proposed model optimally schedules microgrid resources for supporting the distribution grid's flexibility requirements. These flexibility requirements were considered in terms of net load ramping limits. The model was studied for intra-hour and inter-hour time intervals during the 24-hour day-ahead operation. The Θ-k islanding criterion was further taken into account to ensure that the microgrid has the capability to switch to the islanded mode, if needed, while supporting the utility grid during the grid-connected operation. A set of case studies was conducted on a test distribution feeder with one microgrid and several consumers and prosumers to demonstrate the effectiveness of the proposed model. The results showed that the grid operator can efficiently leverage the flexibility of existing microgrids in distribution networks to address some of the most pressing flexibility-associated challenges while eliminating the need for costly investments in the generation and distribution facilities.

References

1 U.S. Energy Information Administration, "Annual Energy Review," [Online]. Available: https://www.eia.gov/outlooks/aeo/, 2023.

2 G. Barbose, N. Darghouth, and S. Weaver, Lawrence Barkley National Laboratory (LBNL), "Tracking the Sun VII: An Historical Summary of the Installed Price of Photovoltaics in the United States from 1998 to 2013," Sept. 2014.

3 California ISO, "What the Duck Curve Tells Us About Managing a Green Grid,". [Online]. Available: https://www.caiso.com/Documents/ FlexibleResourcesHelpRenewables_FastFacts.pdf, 2016.

4 A. Costa, A. Crespo, J. Navarro, G. Lizcano, H. Madsen, and E. Feitosa, "A review on the young history of the wind power short-term prediction," *Renew. Sustain. Energy Rev.*, vol. 12, no. 6, pp. 1725–1744, 2008.

5 A. Ulbig and G. Andersson, "On operational flexibility in power systems," in *Proc. 2012 IEEE Power and Energy Society General Meeting*, San Diego, CA, pp. 1–8, 2012.

6 E. Lannoye, D. Flynn, and M. O'Malley, "Evaluation of power system flexibility," *IEEE Trans. Power Syst.*, vol. 27, no. 2, pp. 922–931, 2012.

7 D. S. Kirschen, A. Rosso, J. Ma, and L. F. Ochoa, "Flexibility from the demand side," in *Proc. IEEE Power and Energy Society General Meeting*, San Diego, CA, pp. 1–6, 22–26 July 2012.

8 M. Ghamkhari and H. Mohsenian-Rad, "Optimal integration of renewable energy resources in data centers with behind-the-meter renewable generator," in *Proc. IEEE International Conference on Communications (ICC)*, Ottawa, ON, pp. 3340–3344, 10–15 June 2012.

9 A. Asadinejad and K. Tomsovic "Impact of incentive based demand response on large scale renewable integration," in *2016 IEEE PES Innovative Smart Grid Technologies (ISGT)*, Minneapolis, MN, 6–9 Sept. 2016.

10 S. C. Chan, K. M. Tsui, H. C. Wu, Y. Hou, Y. Wu, and F. F. Wu, "Load/price forecasting and managing demand response for smart grids: methodologies and challenges," *IEEE Signal Process. Mag.*, vol. 29, no. 5, pp. 68–85, 2012.

11 O. Ma, N. Alkadi, P. Cappers, P. Denholm, J. Dudley, S. Goli, M. Hummon, S. Kiliccote, J. MacDonald, N. Matson, D. Olsen, C. Rose, M.D. Sohn, M. Starke, B. Kirby, and M. O'Malley, "Demand response for ancillary services," *IEEE Trans. Smart Grid*, vol. 4, no. 4, pp. 1988–1995, 2013.

12 J. Aghaei and M. I. Alizadeh, "Demand response in smart electricity grids equipped with renewable energy sources: a review," *Renew. Sustain. Energy Rev.*, vol. 18, pp. 64–72, 2013.

13 P. Denholm, E. Ela, B. Kirby, and M. Milligan, "The Role of Energy Storage With Renewable Electricity Generation," National Renewable Energy Laboratory, Tech. Rep. NREL/TP-6A2-47187, Available: http://www.nrel.gov/docs/fy10osti/47187.pdf, Jan. 2010.

14 J. Eyer and G. Corey, "Energy Storage for the Electricity Grid: Benefits and Market Potential Assessment Guide; A Study for the DOE Energy Storage Systems Program," Sandia National Laboratories, 2010.

15 M. Beaudin, H. Zareipour, A. Schellenberglabe, and W. Rosehart, "Energy storage for mitigating the variability of renewable electricity sources: an updated review," *Energy Sustain. Dev.*, vol. 14, no. 4, pp. 302–314, 2010.

16 P. Denholm and M. Hand, "Grid flexibility and storage required to achieve very high penetration of variable renewable electricity," *Energy Policy*, vol. 39, no. 3, pp. 1817–1830, 2011.

17 K. Bradbury, L. Pratson, and D. Patino-Echeverri, "Economic viability of energy storage systems based on price arbitrage potential in real-time U.S. electricity markets," *Appl. Energy*, vol. 114, pp. 512–519, 2014.

18 S. Gottwalt, A. Schuller, C. Flath, H. Schmeck, and C. Weinhardt, "Assessing load flexibility in smart grids: electric vehicles for renewable energy integration," in *Proc. 2013 IEEE Power and Energy Society General Meeting*, Vancouver, BC, pp. 1–5, 21–25 July 2013.

19 S.-I. Image, "Modelling Load Shifting Using Electric Vehicles in a Smart Grid Environment," IEA/OECD Working Paper. [Online]. Available: http://www.iea. org/publications/freepublications/publication/load_shifting.pdf, 2010.

20 A. Majzoobi and A. Khodaei, "Application of microgrids in addressing distribution network net-load ramping," in *IEEE PES Innovative Smart Grid Technologies Conference (ISGT)*, Minneapolis, MN, 6–9 September 2016.

21 Department of Energy Office of Electricity Delivery and Energy Reliability, "Summary Report: 2012 DOE Microgrid Workshop," [Online]. Available: http:// energy.gov/sites/prod/files/2012MicrogridWorkshopReport09102012.pdf, 2012.

22 A. Majzoobi and A. Khodaei "Leveraging Microgrids for Capturing Uncertain Distribution Network Net Load Ramping," in *North American Power Symposium (NAPS)*, Denver, CO, 18–20 September 2016.

23 A. Khodaei, "Microgrid optimal scheduling with multi-period islanding constraints," *IEEE Trans. Power Syst.*, vol. 29, no. 3, pp. 1383–1392, 2014.

24 A. Khodaei, "Resiliency-oriented microgrid optimal scheduling," *IEEE Trans. Smart Grid*, vol. 5, no. 4, pp. 1584–1591, 2014.

25 A. Majzoobi, A. Khodaei, S. Bahramirad, and M. H. J. Bollen, "Capturing the variabilities of distribution network net-load via available flexibility of microgrids," in *CIGRE Grid of the Future Symposium*, Philadelphia, PA, 30 Oct.–1 Nov. 2016.

26 A. Khodaei, "Provisional microgrid planning," *IEEE Trans. Smart Grid*, vol. 8, no. 3, pp. 1096–1104, 2015.

27 F. Wu, X. Li, F. Feng, and H. B. Gooi, "Modified cascaded multilevel grid-connected inverter to enhance European efficiency," *IEEE Trans. Industr. Inform.*, vol. 11, no. 6, pp. 1358–1365, 2015.

28 M. J. Sanjari, A. H. Yatim, and G. B. Gharehpetian, "Online dynamic security assessment of microgrids before intentional islanding occurrence," *Neural Comput. Appl.*, vol. 26, no. 3, pp. 659–668, 2014.

29 M. Shahidehpour and J. F. Clair, "A functional microgrid for enhancing reliability, sustainability, and energy efficiency," *Electr. J.*, vol. 25, no. 8, pp. 21–28, 2012.

30 Microgrid Exchange Group, "DOE Microgrid Workshop Report," [Online]. Available: http://energy.gov/oe/downloads/microgrid-workshop-report-august-2011, 2011.

31 Navigant Research, "Microgrid Deployment Tracker 4Q13; Commercial/Industrial, Community/Utility, Institutional/Campus, Military, and Remote Microgrids: Operating, Planned, and Proposed Projects by World Region," 2013.

32 S. Parhizi, H. Lotfi, A. Khodaei, and S. Bahramirad, "State of the art in research on microgrids: a review," *IEEE Access*, vol. 3, pp. 890–925, 2015.

33 W. Pei, Y. Du, W. Deng, K. Sheng, H. Xiao, and H. Qu, "Optimal bidding strategy and intramarket mechanism of microgrid aggregator in real-time balancing market," *IEEE Trans. Industr. Inform.*, vol. 12, no. 2, pp. 587–596, 2016.

34 S. I. Vagropoulos, S. Member, A. G. Bakirtzis, and S. Member, "Optimal bidding strategy for electric vehicle aggregators in electricity," *IEEE Trans. Power Syst.*, vol. 28, no. 4, pp. 4031–4041, 2013.

35 H. Kim and M. Thottan, "A two-stage market model for microgrid power transactions via aggregators," *Bell Labs Tech. J.*, vol. 16, pp. 101–107, 2011.

36 D. T. Nguyen and L. B. Le, "Risk-constrained profit maximization for microgrid aggregators with demand response," *IEEE Trans. Smart Grid*, vol. 6, no. 1, pp. 135–146, 2015.

37 M. J. Sanjari, H. Karami, and H. B. Gooi, "Micro-generation dispatch in a smart residential multi-carrier energy system considering demand forecast error," *Energ. Conver. Manage.*, vol. 120, pp. 90–99, 2016.

38 L. Shi, Y. Luo, and G. Y. Tu, "Bidding strategy of microgrid with consideration of uncertainty for participating in power market," *Electr. Power Energy Syst.*, vol. 59, pp. 1–13, 2014.

39 Y. Gu and L. Xie, "Stochastic look-ahead economic dispatch with variable generation resources," *IEEE Trans. Power Syst.*, vol. 32, no. 1, pp. 17–29, 2016.

40 B. A. Frew, S. Becker, M. J. Dvorak, G. B. Andersen, and M. Z. Jacobson, "Flexibility mechanisms and pathways to a highly renewable US electricity future," *Energy*, vol. 101, pp. 65–78, 2016.

41 Y. Dvorkin, M. A. Ortega-Vazquez, and D. S. Kirschen, "Assessing flexibility requirements in power systems," *IET Gener. Transm. Distrib.*, vol. 8, no. 11, pp. 1820–1830, 2014.

42 J. Ma, "Evaluating and planning flexibility in a sustainable power system with large wind penetration," Ph.D. dissertation, Dept. Electrical and Electronic Eng., Univ. Manchester, Manchester, 2012

43 Estimating the Value of Lost Load [Online]. Available: http://www.ercot.com/content/gridinfo/resource/2015/mktanalysis/ERCOT_ValueofLostLoad_LiteratureReviewandMacroeconomic.pdf.

11

Microgrid Operations Under Electricity Market Dynamics

11.1 Principles of Microgrid Operations Under Electricity Markets

The microgrid capability to operate as a single controllable entity, which is enabled through application of dispatchable distributed energy resources (DERs) and flexible loads, allows for an active participation in a variety of demand response programs in response to economic and emergency incentives [1–9].

Establishing efficient operations and control is one of the most challenging tasks in managing microgrids. The microgrid control is commonly performed in three hierarchical levels, that is primary, secondary, and tertiary [10, 11]. The first two control levels deal with droop control and frequency/voltage adjustment and restoration when there is a change in the microgrid load and/or generation as well as in islanding transitions. The third control level schedules microgrid components and determines the interactions with the main grid while taking the economic and reliability aspects into consideration. Microgrid scheduling problem aims to minimize the operations costs of local DERs, as well as the energy exchange with the main grid, to supply forecasted local loads in a given time period (typically one day). This problem is subject to a variety of operational constraints, such as power balance and DER limitations, and is performed by the microgrid controller. Extensive discussions are available in the literature on the architecture of the microgrid controller, including decentralized [12, 13], centralized [14–16], and hybrid microgrids [17, 18], and also on the methods to solve the scheduling problem with a primary focus on accurate component modeling and uncertainty consideration, such as deterministic methods [19–21], stochastic programming [22, 23], chance-constrained [24], and robust optimization [12]. Additional discussions can be found in the literature, solving the problem using multi-agent systems [25], or benefiting from heuristic methods such as particle swarm optimization [26].

The Economics of Microgrids, First Edition. Amin Khodaei and Ali Arabnya.
© 2024 The Institute of Electrical and Electronics Engineers, Inc.
Published 2024 by John Wiley & Sons, Inc.

Increasing demand-side elasticity and active participation of loads in the power system, commonly in response to electricity price variations, is highly stressed to operate the system more efficiently and to avoid high price spikes caused by inelastic loads [27]. Microgrids allow an efficient integration and control of large penetration of responsive loads, which would further increase the demand-side elasticity. Moreover, distributed generators (DGs) and energy storage support a relatively fast and highly controllable load. However, these resources are typically scheduled based on a price-based scheduling model, i.e. the microgrid controller determines the least-cost schedule of available DERs and loads, as well as the main grid power transfer, based on the day-ahead market prices – which are forecasted by the microgrid or the electric utility. Under this scheme, the utility forecasts an estimate of the microgrids' loads in its service territory and submits it to the system operator. Once the price of electricity is determined, through the wholesale market, the utility sends the actual prices to microgrids. Although it might seem efficient, this approach has the potential to cause several drawbacks when the microgrid penetration in distribution network is high, including but not limited to shifting the peak hours. This approach is prone to cause new peaks since there is a high probability that microgrids follow a different schedule compared to the one forecasted by the utility once actual prices are received, considering that the power demand in responsive loads is inversely proportional to electricity prices. The increase in the number of entities with responsive loads operated based on price-based scheme would intensify this issue. In other words, setting the price centrally by the system operator and sending it to microgrids, so they can accordingly schedule their resources, can potentially result in significant uncertainty in the system load profile. The increased penetration of DERs and microgrids would also make it more challenging to ensure distribution system reliability and desired service level [28].

The concept of aggregators is one of the ideas that was proposed to address these issues. Aggregator discussions can be found in [29], where it is proposed to iteratively collect power generated by microgrids, sell this power to the main grid, and accordingly gain profit via a price-based scheduling. In [30], an aggregator for electric vehicles with fixed energy costs is proposed. The study in [31] proposes a framework for interactions between the customers in a distribution system as well as the main grid, while [32] proposes an entity between the market operator and customers that compensates the aggregators for the services they provide. A coupon incentive-based demand response model is further proposed in [33] enabling customers to increase their flexibility and lower their costs. The proposed model in [34] enables a demand response aggregator to participate in the electricity market, considering market price to be constant, and is further applied to microgrids in [35].

Challenges described earlier combined with the increasing complexity in managing a large number of microgrids in the near future have made the case for developing new methodologies for the system operation and utility ratemaking in presence of microgrids. The concept of a Distribution System Operator (DSO) is proposed as an entity, which is hosted in the distribution network to manage interaction of microgrids with the main grid. In line with the ongoing trend in proposing electricity markets in distribution networks, we propose a market-based microgrid optimal scheduling model to address the aforementioned problems and increase microgrid-integrated distribution system efficiency and social welfare. Considering that a DSO offers both grid and market functionalities, we only focus on the market operations and provide discussions on Distribution Market Operator (DMO) concept. The DMO can be considered as the distribution level equivalent of the Independent System Operator (ISO), which is responsible for managing the electricity market and scheduling power transfers to achieve the optimal operation in the distribution network [36]. Considering limited studies on the viability of distribution markets, this chapter aims to: (i) discuss the necessity of the DMO in future power grids and identify interactions with connected market players, (ii) formulate the three levels of the market structure, i.e. ISO, utility, and customer levels, to provide an insight on the data exchange and involved optimization problems in these levels, (iii) develop an analytical model for the market-based microgrid scheduling, and formulate the problem using mixed-integer linear programming, and (iv) use the developed comprehensive model to enable comparisons between price-based and market-based scheduling schemes from a microgrid and a system-level perspectives.

One may argue that a highly accurate load estimation would resolve the aforementioned issues, in which the system operators would have a fairly accurate idea of load fluctuations. However, it should be noted that the benefits of the DSO are not limited to improving predictability. The DSO can provide local resilience capability [28] and can reduce dependence on the ISO for providing balancing services, so the distribution system can maintain its service when the rest of the system is in an abnormal condition [37]. It can also manage the energy transactions between the DERs and the loads within the distribution system; demand for this service would grow as the number of such transactions increases [28]. State of New York's Reforming The Energy Vision (REV) strategy asserts that in order to "create a more robust retail market" it is necessary to provide market operations and grid operations at the distribution level [38]. Easing complexity of direct scheduling of responsive loads and DERs in the wholesale market, solving scalability issues and providing ancillary services are among other beneficial functionalities that the DSO can provide to the distribution system [37–39]. On the other hand, the ISOs may not have control over the demand side assets, so those assets need to have the capability to provide the required reserve and flexibility services to handle variable resources [40]. In [41], a price-based simultaneous operation of microgrids and the Distribution Network Operator (DNO) is proposed. In the State of New York, the new concept of Distributed System Platform Provider (DSPP) was introduced as part of the REV program [38] where the transformation of existing utility operations to integrate high penetration of microgrids and DERs is discussed. DSPPs can be formed as new entities or be part of the existing electric utilities. An independent DSPP would be able to set up a universal market environment instead of one for each utility. It would also be less suspected of exercising market power. A utility-affiliated DSPP, however, would be able to perform several functionalities currently possessed by electric utilities without necessitating additional investments. In California, the state public utilities commission has ruled to establish regulations to guide investor-owned electric utilities in developing their Distribution Resources Plan proposals. Studies in [28, 42] provide a framework for this ruling, defined as a DSO, which is in charge of operation of local distribution area and providing distribution services. The DSO is further responsible to provide forecasting and measurement to the ISO and manage power flow across the distribution system. It is also suggested that the DSO adopt further roles such as coordination of dispersed units in the distribution network and providing an aggregate bid to the ISO. The study in [40] proposes a DSO as an ISO for the distribution network, which is responsible for balancing supply and demand at the distribution level, linking wholesale and retail market agents, and linking the ISO to the demand side. In contrast to the European definition of the DSO, the proposed entity in [40] interacts directly with the ISO. The study further presents a spectrum of different levels of the DSO autonomy in operating the distribution system and the ISO's degree of control over it. From the least degree of autonomy to the most degree of autonomy, this spectrum entails the DSO to be able to perform the forecasting and send it to the ISO, be responsible for balancing the supply and demand, be able to receive offers from DERs, aggregate them and bid into the wholesale market, and eventually be able to control the retail market so that various DERs can have transactions not only with the DSO but among themselves. In [43], an independent distribution system operator (IDSO) is proposed to be responsible for distribution grid operations, while grid ownership remains in the hands of utilities. The IDSO is envisioned to provide market mechanisms in the distribution system, enable open access, and ensure safe and reliable electricity service. The IDSO will reduce the operational burden on utilities and determine the true value of resources in a more objective manner. The importance of distribution markets in integrating proactive customers has been emphasized in the literature [44–46].

11.2 Economic Variables of Electricity Markets in Microgrid Operations

As discussed earlier, DMO is a platform that enables market activities for end-use customers, coordinates with the utility to improve grid operations, and interacts with the ISO to determine demand bid awards. The DMO will further facilitate establishing a competitive electricity market in the distribution network to exchange energy and grid services with customers and expedite a more widespread integration of DERs from a system operator's perspective by addressing prevailing integration challenges. Figure 11.1 depicts the interactions of different players in the market in presence of the DMO at three levels, that is ISO, utility, and customer levels.

Two major responsibilities of the DMO within this structure are, as follows: (i) to receive demand bids from the microgrids (and other responsive loads if any), combine them, and offer an aggregated bid to the ISO and (ii) to receive the day-ahead schedule from the ISO, solve a resource scheduling problem for its service territory, and subsequently determine microgrids' shares from the awarded power.

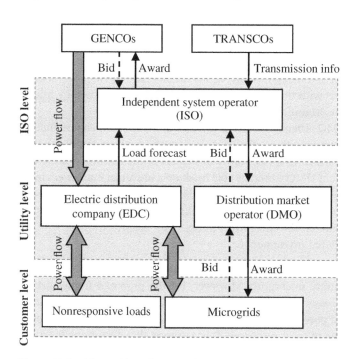

Figure 11.1 Microgrid market participation through the proposed DMO.

Microgrids would submit their bids to the DMO (in the form of monotonically decreasing demand bids) and later be notified by the DMO on the amount of awarded power (henceforth referred to as the assigned power). Other responsive loads can be considered at the customer level without loss of generality. The main grid power transfer to the microgrid would be the amount of power assigned to the microgrid by the DMO, hence it would be known to the system operator in advance and therefore eliminate the uncertainties caused by microgrids to a large extent. Once the main grid power transfer is reported to the microgrid, for the 24 hours of the next day, the microgrid would solve a market-based scheduling problem to optimally schedule its DERs and loads. Only private microgrids are considered here as there could be some regulatory barriers in market participation of utility-owned microgrids depending on their jurisdictions. This framework offers several advantages, as follows:

- The microgrid demand is set by the DMO and known with certainty on a day-ahead basis. This will lead to manageable peak demands and increased operational reliability and efficiency. Microgrids will have the capability to deviate from the assigned power (as it will be further discussed later in this chapter), however, it will be at the expense of paying a penalty, hence potential deviations would be minimal.
- The microgrid can exchange power with the main grid and act as a player in the electricity market. The DMO would serve as an intermediate entity between the ISO and microgrids that facilitates microgrids market participation and coordinates the microgrids with the main grid to minimize the risks posed by microgrid operational uncertainties.
- Establishing the DMO is beneficial to the ISO as it allows a significant reduction in the required communication infrastructure between microgrids and the ISO.
- The DMO can be formed as a new entity or be part of the existing electric utilities. An independent DMO would be able to set up a universal market environment instead of one for each utility. It would also be less suspected of exercising market power. On the other hand, a utility-affiliated DMO would be able to perform several functionalities currently possessed by electric utilities without necessitating additional investments.

Implementation of the DMO would fix the aforementioned issues that utilities face when they integrate microgrids. However, for the proposed framework to work reliably, it is necessary that the microgrid controller to schedule its resources based on the assigned power transfer, considering that microgrid controller seeks the least-cost schedule of local resources. It is assumed that deviations from the assigned value will be penalized based on the market price or a relatively larger value that can effectively prevent and/or reduce deviations. It is further assumed that the penalty will be applied when the deviation is positive, i.e. the microgrid's scheduled power is larger than the assigned power by the DMO, or in other words, when the microgrid appears as a larger load compared to the assigned power by the DMO. Negative deviation will not be penalized in the proposed model as the microgrid helps with reducing load (increasing generation) in the distribution network. This issue is further investigated in case studies in this chapter.

Under the price-based scheduling method, the ISO receives load forecasts from Electric Distribution Companies (EDC) and determines the day-ahead unit schedules by solving a security-constrained unit commitment (SCUC) problem. The ISO will also determine locational marginal prices (LMPs), which will be further used by microgrids for scheduling purposes. In the price-based method, contrary to the market-based method, there is no need for microgrids to offer bids and participate in the market, and moreover, the main grid power transfer will be determined via a local cost minimization problem rather than being determined by the DMO via a market mechanism. In either method, however, the microgrid needs to determine the optimal schedule of local DERs and loads to address its energy needs and ensure a reliable supply of local loads.

11.3 An Economic Model for Microgrid Operations Planning Under Market Dynamics

The three levels of the discussed market structure are formulated in the following to provide an insight on the data exchange, represent optimization problems involved in different levels, and further enable numerical studies on microgrid scheduling.

11.3.1 Microgrid Level

Microgrids determine the least-cost, day-ahead schedule of their loads, dispatchable generation units, and energy storage considering a known profile for the main grid power transfer, which is determined and assigned by the DMO, over the planning horizon. Each microgrid m solves the proposed market-based optimal scheduling problem through (11.1)–(11.10), as follows:

$$\min \sum_t \sum_i \left[F_{im}(P_{imt}, I_{imt}) + \upsilon_m LS_{mt} + \nu_m \Delta P_{mt}^{M+} \right] \tag{11.1}$$

s.t.

$$P_{mt}^M + \sum_i P_{imt} + LS_{mt} = d_{mt} \qquad \forall t \tag{11.2}$$

$$P_{im}^{\min} I_{imt} \leq P_{imt} \leq P_{im}^{\max} I_{imt} \qquad \forall t, \forall i \tag{11.3}$$

$$\sum_t P_{imt} = E_{im} \qquad \forall t \tag{11.4}$$

$$\sum_t f_{im}(P_{imt}, I_{imt}) \leq 0 \qquad \forall i \tag{11.5}$$

$$-P_m^{M,\max} U_{mt} \leq P_{mt}^M \leq P_m^{M,\max} U_{mt} \qquad \forall t \tag{11.6}$$

$$\Delta P_{mt}^M = P_{mt}^M - PD_{mt}^M \qquad \forall t \tag{11.7}$$

$$-P^{M,\max} \delta_{mt} \leq \Delta P_{mt}^{M+} \leq P^{M,\max} \delta_{mt} \qquad \forall t \tag{11.8}$$

$$-P^{M,\max}(1 - \delta_{mt}) \leq \Delta P_{mt}^M - \Delta P_{mt}^{M+} \leq P^{M,\max}(1 - \delta_{mt}) \qquad \forall t \tag{11.9}$$

$$-P^{M,\max}(1 - \delta_{mt}) + \varepsilon \leq \Delta P_{mt}^M \leq P^{M,\max} \delta_{mt} \qquad \forall t \tag{11.10}$$

where b is index for buses, i is index for DERs, t is index for hours, m is index for microgrids, P is variable for DER output power, I is variable for commitment state of dispatchable unit (1 when committed, 0 otherwise), LS is variable for load curtailment, ΔP^{M+} is a variable for positive power transfer deviation, P^M is variable for scheduled power transfer from the DMO to the microgrid, d is variable for load demand, $f(P,I)$ is the generation cost function of the dispatchable unit, ΔP^M is a variable for power transfer deviation, PD^M is a variable for assigned demand to microgrids by the DMO, δ is a variable for power transfer deviation indicator (1 when deviation is positive, 0 otherwise), ν is a parameter for penalty for scheduled power deviation, v is the value of lost load (VOLL), and U is a parameter for islanding indicator (1 when grid-connected, 0 when islanded). The three terms in the objective function (11.1) represent the operation cost, the load curtailment cost, and the deviation cost, respectively. The operations cost is the cost of power production as well as startup and shutdown costs of dispatchable units. The load curtailment cost is defined as the VOLL times the amount of load curtailment. The VOLL is assumed as an opportunity cost based on the cost that the consumer is willing to pay to have reliable uninterrupted service. The VOLL is commonly used as a measure of load criticality [47]. The deviation cost is the penalty imposed on the microgrid in case the microgrid schedule deviates from the power transfer assigned by the DMO. The objective is subject to a set of operational constraints (11.2)–(11.10). The power balance Eq. (11.2) ensures that the sum of the main grid power plus the locally generated power from DERs matches the total load, while load curtailment variable is added to ensure that the power balance is satisfied at all times. Non-dispatchable generation and fixed load values are assumed to be forecasted with acceptable accuracy and are treated as uncontrollable parameters. There of course would be uncertainties associated with possible forecast errors. All operational constraints associated with DERs and loads are formulated using three general constraints (11.3)–(11.5), respectively, representing power constraints, energy constraints, and time-coupling constraints. Power constraints (11.3) account for power capacity limits, such as dispatchable generation minimum/ maximum capacity limits, energy storage minimum/maximum charge/discharge power, and flexible load minimum/maximum capacity limits. Energy constraints (11.4) account for energy characteristics of a specific DER or load, such as energy storage state of charge limit and flexible load required energy in a cycle. Time-coupling constraints (11.5) represent any constraint that links variables in two or more scheduling hours, such as dispatchable units ramp up/down rates and minimum on/off times, energy storage rate and profile of charge/discharge, and adjustable loads minimum operating time and load pickup/drop rates. Using these constraints, any type of DER and load can be efficiently modeled. A detailed modeling of microgrid DERs and loads can be found in [48]. The main grid power transfer is restricted by its associated limits (imposed by the capacity of the line connecting the microgrid to the main grid) in (11.6). The islanding is modeled using a binary islanding indicator U, which would zero out the main grid power transfer when 0. The main grid power deviation to be penalized in the objective is determined in (11.7)–(11.10). Constraint (11.7) calculates the deviation by subtracting the scheduled power via the optimal scheduling, P^M, by the assigned power from the DMO, PD^M. Constraints (11.8)–(11.10) determine the penalty if the calculated deviation is positive. An auxiliary binary variable δ is used for this purpose. When $\delta = 0$ the power transfer to be penalized is zero, i.e. the scheduled power is less than the assigned power. However, when $\delta = 1$ the power transfer to be penalized is equal to the positive deviation calculated in (11.7).

11.3.2 DMO Level

The DMO seeks two objectives, as follows: first, to combine individual bids received from microgrids in its territory to create an aggregated bid and accordingly send the aggregated bid to the ISO to participate in the energy market; second, to disaggregate the awarded quantity by the ISO to individual microgrids in accordance with their respective bids. These two tasks are discussed in the following.

Bid aggregation: Figure 11.2 depicts a typical demand bid curve submitted by a microgrid to the DMO at a given hour t. The bid consists of fixed and variable parts. The fixed part shows the microgrid nonresponsive load, which must be fully supplied under normal operation conditions and cannot be altered. The variable part, on the other hand, shows the microgrid flexibility in reducing its consumption from its total load. It consists of several segments. The reduction in consumption can be achieved either via load curtailment or local DER generation. The DMO combines the individual microgrid bids and obtains an aggregated bid to be sent

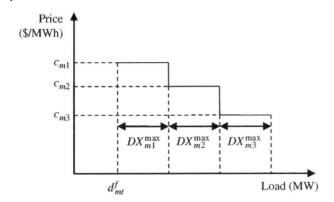

Figure 11.2 Demand bid curve for microgrid m with a three-segment bid.

to the ISO. The fixed loads are collectively added to obtain the total fixed load in the DMO service territory, as follows:

$$D_t^f = \sum_m d_{mt}^f \qquad \forall t \tag{11.11}$$

where f is superscript for fixed loads, D^f is a parameter for total fixed load of all microgrids in the distribution network, and d^f is a variable for fixed load demand.

Quantity disaggregation: Once the ISO determines the awarded power to the DMO, the DMO disaggregates the power to microgrids in its service territory. The DMO maximizes the objective function (11.12) by determining the optimal allocated power to each microgrid based on the submitted bids, as follows:

$$\max \sum_t \sum_m \sum_j c_{mj} DX_{mjt} \tag{11.12}$$

s.t.

$$DX_{mjt} \le DX_{mj}^{\max} \qquad \forall m, \forall t, \forall j \tag{11.13}$$

$$d_{mt}^r = \sum_j DX_{mjt} \qquad \forall m, \forall t \tag{11.14}$$

$$d_{mt}^f + d_{mt}^r = PD_{mt}^M \qquad \forall m, \forall t \tag{11.15}$$

$$\sum_m PD_{mt}^M = D_{bt} \qquad \forall t \tag{11.16}$$

where j is index for segments of the load bids, b is index for buses, r is superscript for responsive loads, DX is a variable for the amount of load awarded to each segment of the bid, $DX^{\max}$ is a variable for the segment in the variable load bid of the microgrid, d^r is a variable for responsive load demand, and D is the demand awarded from the ISO to the DMO. Constraint (11.13) guarantees that each segment of load is limited by its maximum. The total responsive demand for each

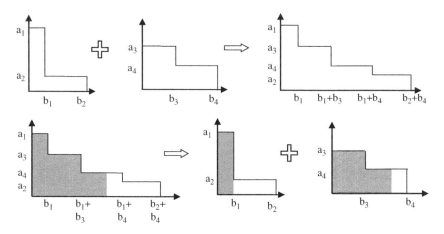

Figure 11.3 An example of DMO aggregating two submitted bids (top) and disaggregating awarded power (bottom). Vertical and horizontal axes show price and load, respectively.

microgrid is the sum of the loads dispatched to each associated segment (11.14). The awarded load is calculated as the summation of the fixed and responsive loads (11.15), and accordingly, the amount of power flow from the ISO to the DMO as the summation of the awarded loads is set by Eq. (11.16) as the total load dispatched to all load segments is equal to the assigned power by the ISO. Figure 11.3 provides a graphical representation of the bid aggregation and quantity disaggregation by the DMO. The distribution line limits in this model are assumed to be adequately large to handle any power transfer without causing congestion in the distribution network. Additional constraints, however, can be simply added to the model, including but not limited to distribution line power flow and limits, ramp rate constraints, etc. Another important constraint that can be considered is the load-shifting capability of microgrids. Modeling the load shifting would require the inclusion of time-coupling constraints among hourly bids. This topic is addressed for ISOs in previous work of authors [49].

11.3.3 ISO Level

The ISO receives the generation and transmission information, respectively, from generation companies (GENCOs) and transmission companies (TRANSCOs), receives the demand bids from DMOs, and solves the SCUC problem to determine units schedule followed by a security-constrained optimal power flow to determine unit dispatch, line flow, and LMPs. The ISO's objective, when considering demand bids, will be to maximize the system social welfare, rather than minimizing the total operations cost (11.17) under a set of constraints (11.18)–(11.23), as follows:

$$\max\left\{\sum_t\sum_b \lambda_{bt}(D_{bt}) - \sum_t\sum_i \rho_{it}(P_{it})\right\} \tag{11.17}$$

s.t.

$$\sum_{i \in G_b} P_{it} - \sum_{l \in L_b} PL_{lt} = D_{bt} \qquad \forall t, \forall b \tag{11.18}$$

$$P_i^{\min} I_{it} \le P_{it} \le P_i^{\max} I_{it} \qquad \forall t, \forall i \tag{11.19}$$

$$\sum_t P_{it} = E_i \qquad \forall t \tag{11.20}$$

$$\sum_t f_i(P_{it}, I_{it}) \le 0 \qquad \forall t, \forall i \tag{11.21}$$

$$|PL_{lt}| \le PL_l^{\max} \qquad \forall t, \forall l \tag{11.22}$$

$$PL_{lt} = \sum_b \frac{B_{lb}\theta_{bt}}{x_l} \qquad \forall t, \forall l \tag{11.23}$$

where l is index for transmission lines, PL is variable for line power flow, θ is variable for bus angle, $\rho(P)$ is a parameter for cost function of generation units submitted to the ISO, $\lambda(D)$ is a parameter for consumption benefit of aggregated loads, B is parameter for components of the bus-to-line incidence matrix, $PL^{\max}$ is a parameter for line flow limit, and x is a parameter for line impedance. The ISO maximizes objective function (11.17), which is a system-level social welfare, i.e. consumption payments minus generation costs. This objective function is subject to the power balance constraint (11.18), unit commitment constraints (11.19)–(11.21), transmission line limits (11.22), and transmission line power flow (11.23). Unit commitment constraints include unit output limits, unit spinning/operating reserve limit, ramp up/down rate limits, min up/downtime limits, fuel limits, and emission limits. More details on the SCUC models can be found in [49].

11.4 Case Study

The proposed market-based microgrid scheduling model is numerically analyzed and compared with the price-based scheduling using the IEEE 118-bus standard test system shown in Figure 11.4. A total of five microgrids are considered to be connected to bus 60 with a total installed DG capacity of 50 MW, which is equal to 51% of the peak load at this bus. The specifications of microgrid DGs are given in Table 11.1. Specifications of adjustable loads, energy storage, and fixed loads are borrowed from [48].

Two cases including a price-based microgrid optimal scheduling and a market-based microgrid optimal scheduling are considered, as follows.

Case 1: In this case, the ISO uses the forecasted microgrid loads to clear the market and accordingly determine hourly LMP values. Microgrids individually

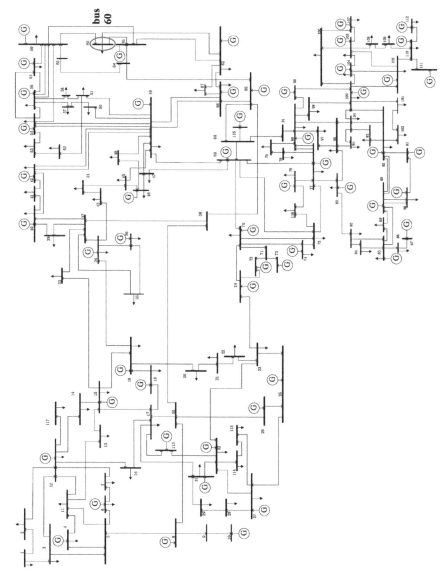

Figure 11.4 IEEE 118-bus standard test system.

Table 11.1 Cost characteristics of DG units.

	Capacity (MW)				
	MG1	**MG2**	**MG3**	**MG4**	**MG5**
DG1	4	5	3	4	3
DG2	3	3	3	3	3
DG3	2	1	2	2	2
DG4	1	1	2	1	2
	Price ($/MWh)				
DG1	27.5	29.1	27.4	28.3	33.5
DG2	43.1	37.3	38.2	35.3	41.1
DG3	64.3	59.3	55.2	60.3	65.5
DG4	69.6	70.9	61.1	62.4	72.2

perform their own scheduling using the LMP values. With microgrids being connected to bus 60, five lines in the system, including one of those connected to the bus 60, become congested at peak hours. The total microgrid operation cost is calculated as $74,447. In this case, the actual amount of load at the bus to which microgrids are connected will not match the amount originally considered by the ISO when clearing the wholesale market. If the ISO runs the economic dispatch with the actual microgrids net loads, which would be less than the microgrid load used initially by the ISO to determine the LMPs, the prices will change. This change in LMPs is a major drawback in the price-based model where there is a mutual and uncontrolled interaction between the calculated LMPs and microgrids net load. Another drawback is the mismatch between the initially forecasted load and the actual load, after microgrid optimal scheduling, which needs to be addressed by the ISO by re-dispatching the committed units. The re-dispatch will potentially result in an economic loss for the system as the new solution will diverge from the already determined optimal dispatch solution.

Case 2: In this case, the bid that each microgrid sends to the DMO is created based on the capacity and marginal costs of its dispatchable DGs. For example, microgrid 2 will have a four-step bid, as follows: a 1 MW bid at $70.9/MWh, a 1 MW bid at $59.3/MWh, a 3 MW bid at $37.3/MWh, and a 5 MW bid at $29.1/MWh, as derived from Table 11.1. Using this bid, the demand responsiveness of the microgrid is modeled by local generation of dispatchable DGs. The total microgrid operation cost in this case is $48,568, which shows a 34.76% reduction from that of Case 1. Table 11.2 shows the committed DGs in each microgrid, in which bold figures represent changes from the price-based optimal scheduling solution in

Table 11.2 The commitment schedule of microgrid DGs.

Microgrid	DG	1	2	3	4	5	6	7	8	9	10	11	12	13	14	15	16	17	18	19	20	21	22	23	24
MG1	DG1	1	1	1	1	1	0	0	0	1	1	1	1	1	1	1	1	1	1	1	1	1	1	1	1
	DG2	0	0	0	0	0	0	0	0	0	0	0	0	0	0	0	0	0	0	0	0	0	0	0	0
	DG3	0	0	0	0	0	0	0	0	0	0	0	0	0	0	0	0	0	0	0	0	0	0	0	0
	DG4	0	0	0	0	0	0	0	0	0	0	0	0	0	0	0	0	0	0	0	0	0	0	0	0
MG2	DG1	0	0	0	0	1	1	0	0	1	1	1	1	1	1	1	1	1	1	1	1	1	1	1	1
	DG2	0	0	0	0	0	0	0	0	0	0	0	0	0	0	0	0	0	0	0	0	0	0	0	0
	DG3	0	0	0	0	0	0	0	0	0	0	0	0	0	0	0	0	0	0	0	0	0	0	0	0
	DG4	0	0	0	0	0	0	0	0	0	0	0	0	0	0	0	0	0	0	0	0	0	0	0	0
MG3	DG1	1	1	1	1	1	0	0	0	1	1	1	1	1	1	1	1	1	1	1	1	1	1	1	1
	DG2	0	0	0	0	0	0	0	0	0	0	0	0	0	1	1	0	0	0	0	0	0	0	0	0
	DG3	0	0	0	0	0	0	0	0	0	0	0	0	0	0	1	1	0	0	0	0	0	0	0	0
	DG4	0	0	0	0	0	0	0	0	0	0	0	0	0	0	0	0	0	0	0	0	0	0	0	0
MG4	DG1	0	0	0	0	0	1	1	1	1	1	1	1	1	1	1	1	1	1	1	1	1	1	1	1
	DG2	0	0	0	0	0	0	0	0	0	0	0	0	0	0	1	0	0	0	0	0	0	0	0	0
	DG3	0	0	0	0	0	0	0	0	0	0	0	0	0	0	0	0	1	0	0	0	0	0	0	0
	DG4	0	0	0	0	0	0	0	0	0	0	0	0	0	0	0	0	0	0	0	0	0	0	0	0
MG5	DG1	0	0	0	0	0	0	0	1	1	1	0	1	1	1	1	1	1	1	0	0	0	0	0	0
	DG2	0	0	0	0	0	0	0	0	1	1	0	0	0	0	0	0	0	0	0	0	0	0	0	0
	DG3	0	0	0	0	0	0	0	0	0	0	0	0	0	0	0	0	0	0	0	0	0	0	0	0
	DG4	0	0	0	0	0	0	0	0	0	0	0	0	0	0	0	0	0	0	0	0	0	0	0	0

Case 1. This table indicates that many DGs committed in Case 1 are not committed in Case 2. In Case 1, microgrid lowers its power transfer as a response to the market price; therefore, it has to commit more local resources to supply loads. DG1 of each microgrid is the most committed unit in both cases since it has the lowest marginal cost compared to other DGs in the same microgrid.

Figure 11.5 depicts the hourly net load at bus 60 to which microgrids are connected. It is observed that during the early hours, the values of net load in the two cases are close. This is due to the low price of electricity during early hours, when a large portion of the submitted bid from the DMO is awarded by the ISO, resulting in a power transfer close to the total load of the microgrids. In Case 1, the entire demand is supplied by the main grid for the same reason. At hours 8–24, as the electricity price increases, the microgrid loads are partially supplied by local DGs.

Figure 11.6 depicts the hourly LMP at bus 60, i.e. the electricity price for the power transferred to the DMO. Case 2 represents a significantly lower price at peak and close to peak hours, which accordingly results in fewer DG commitments, as it is more economical to purchase power from the main grid, and a lower operations cost. Accordingly, the microgrid net load is increased in this case. Figure 11.7 depicts the average LMP of all buses in the system. The values for market-based model are close to or lower than the values for the price-based model except for hours 13–22. This result advocates that although the market-based scheduling may result in lower LMPs for microgrids, it may not necessarily reduce the system LMP on other network buses. The total system operations cost is reduced from

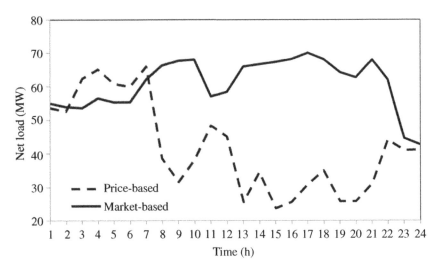

Figure 11.5 Net load at bus 60.

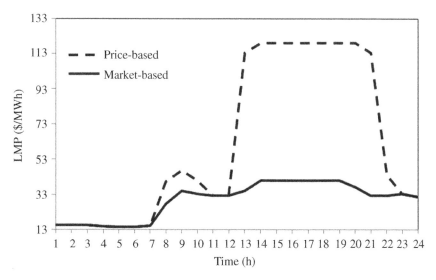

Figure 11.6 LMP at bus 60.

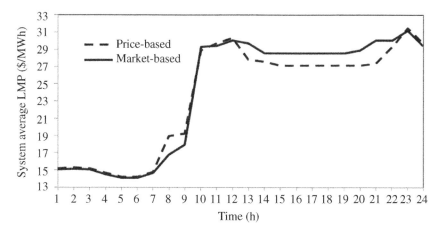

Figure 11.7 Average LMP of 118-bus system.

$1,074,504 in the price-based model to $1,009,734 in the market-based model. To identify the changes in values/trends of LMPs, when such a market is available at all network buses, further investigation is required.

To demonstrate the viability of the proposed deviation reduction method and ensuring that microgrid will follow the DMO-assigned power transfers, the impact of the power transfer deviation penalty is further analyzed. Figure 11.8 depicts the

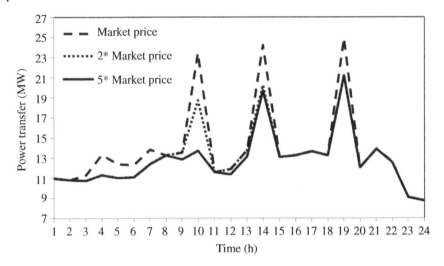

Figure 11.8 Power transfer to microgrid 3 at different levels of deviation penalties.

main grid power transfer to a selected microgrid (MG3) at different levels of deviation penalties. It is assumed that the forecasted microgrid load increases twofold at hours 10, 14, and 19.

Penalties equal to the market price, two times the market price, and five times the market price are considered. The costs of power transfer deviations are, respectively, calculated as $2053, $1777, and $3170. As the penalty increases, the amount of deviation from assigned power decreases but the deviation cost does not linearly change. The microgrid's total operations cost will however decrease. Therefore, it can be seen that higher penalties reduce the amount of deviation to reach the desired values. However, when the penalty becomes too high (comparable to the VOLL) microgrids may economically find it preferable to curtail some loads rather than paying for the penalty in purchasing power from the main grid.

To further demonstrate the impacts of power deviation penalty on the scheduling solutions, two scenarios are considered, as follows: in the first scenario, the microgrid is scheduled based on a price-based scheme after receiving the prices determined by the DMO; in the second scenario, the absolute value of the power deviation, instead of only the positive deviation, is penalized. The microgrid operations cost reduces to $47,380 using price-based scheduling, as the microgrid reduces the power purchase from the main grid at the peak hours and uses its own resources that become price competitive at those times. When the absolute value of the deviations is penalized, the total microgrid operations cost rises to $50,539, as microgrid is obligated to closely follow the scheduled power transfer and hence would reduce generation of some of its resources to purchase more power from the main grid – which results in a higher operations cost. This shows that penalizing power deviation is key in ensuring certainty in the power scheduled by the DMO. Penalizing the absolute value of power deviation can increase the microgrid operations cost, even if the power deviation would result in a surplus of power, which is manageable by the system operator. The decision to penalize only the positive deviations or the absolute deviation should be made by the DSO by incorporating the probable congestion scenarios.

11.5 Summary

In this chapter, we presented an optimal scheduling model for a microgrid participating in the electricity distribution market in interaction with a DMO. The DMO administers an established electricity market at the distribution level, sets electricity prices, determines the amount of the power exchange among market participants, and interacts with the Independent System Operator. Considering a predetermined main grid power transfer to the microgrid, the microgrid scheduling problem aims at balancing the power supply and demand while taking financial objectives into account. Case studies were conducted to demonstrate the application and the effectiveness of the market-based microgrid scheduling model. In addition, the model was benchmarked by comparing it with the commonly used price-based scheduling model to further demonstrate its merits. Results showed that the market-based scheduling model has the potential to lower the microgrid operations cost, lower LMPs at the microgrid-connected buses, and further support system operations by eliminating the net load uncertainty.

References

1 M. Shahidehpour, "Role of smart microgrid in a perfect power system," in *IEEE PES General Meeting*, Minneapolis, MN, p. 1, 2010.
2 A. Khodaei, S. Bahramirad, and M. Shahidehpour, "Microgrid planning under uncertainty," *IEEE Trans. Power Syst.*, vol. 30, no. 5, pp. 2417–2425, 2015.
3 A. Khodaei, "Provisional microgrids," *IEEE Trans. Smart Grid*, vol. 6, no. 3, pp. 1107–1115, 2015.
4 A. Khodaei, "Resiliency-oriented microgrid optimal scheduling," *IEEE Trans. Smart Grid*, vol. 5, no. 4, pp. 1584–1591, 2014.
5 A. Khodaei and M. Shahidehpour, "Microgrid-based co-optimization of generation and transmission planning in power systems," *IEEE Trans. Power Syst.*, vol. 28, no. 2, pp. 1–9, 2012.

6 S. Bahramirad, W. Reder, and A. Khodaei, "Reliability-constrained optimal sizing of energy storage system in a microgrid," *IEEE Trans. Smart Grid*, vol. 3, no. 4, pp. 2056–2062, 2012.

7 V. Cortés de Zea Bermudez, M. Y. El-Sharkh, A. Rahman, and M. S. Alam, "Short term scheduling of multiple grid-parallel PEM fuel cells for microgrid applications," *Int. J. Hydrogen Energy*, vol. 35, no. 20, pp. 11099–11106, 2010.

8 M. Shahidehpour and S. Pullins, "Microgrids, modernization, and rural electrification [about this issue]," *IEEE Electrif. Mag.*, vol. 3, no. 1, pp. 2–6, 2015.

9 S. Parhizi, H. Lotfi, A. Khodaei, and S. Bahramirad, "State of the art in research on microgrids: a review," *IEEE Access*, vol. 3, 2015.

10 M. Shahidehpour and M. Khodayar, "Cutting campus energy costs with hierarchical control: the economical and reliable operation of a microgrid," *IEEE Electrif. Mag.*, vol. 1, no. 1, pp. 40–56, 2013.

11 M. Shahidehpour, "DC microgrids: economic operation and enhancement of resilience by hierarchical control," *IEEE Trans. Smart Grid*, vol. 5, no. 5, pp. 2517–2526, 2014.

12 Y. Zhang, N. Gatsis, and G. B. Giannakis, "Robust energy management for microgrids with high-penetration renewables," *IEEE Trans. Sustain. Energy*, vol. 4, no. 4, pp. 944–953, 2013.

13 N. Cai and J. Mitra, "Economic dispatch in microgrids using multi-agent system," in *2012 North American Power Symposium (NAPS)*, Champaign, IL, pp. 1–5, 2012.

14 R. Enrich, P. Skovron, M. Tolos, and M. Torrent-Moreno, "Microgrid management based on economic and technical criteria," in *2012 IEEE International Energy Conference and Exhibition (ENERGYCON)*, Florence, Italy, pp. 551–556, 2012.

15 C. A. Hernandez-Aramburo, T. C. Green, and N. Mugniot, "Fuel consumption minimization of a microgrid," *IEEE Trans. Ind. Appl.*, vol. 41, no. 3, pp. 673–681, 2005.

16 D. E. Olivares, C. A. Canizares, and M. Kazerani, "A centralized optimal energy management system for microgrids," in *2011 IEEE Power and Energy Society General Meeting*, Detroit, MI, pp. 1–6, 2011.

17 Q. Sun, J. Zhou, J. M. Guerrero, and H. Zhang, "Hybrid three-phase/single-phase microgrid architecture with power management capabilities," *IEEE Trans. Power Electron.*, vol. 30, no. 10, pp. 5964–5977, 2015.

18 N. Eghtedarpour and E. Farjah, "Power control and management in a hybrid AC/DC microgrid," *IEEE Trans. Smart Grid*, vol. 5, no. 3, pp. 1494–1505, 2014.

19 S. Bahramirad and W. Reder, "Islanding applications of energy storage system," in *2012 IEEE Power and Energy Society General Meeting*, San Diego, CA, pp. 1–5, 2012.

20 M. Stadler, A. Siddiqui, C. Marnay, H. Aki, and J. Lai, "Control of greenhouse gas emissions by optimal DER technology investment and energy management in zero-net-energy buildings," *Eur. Trans. Electr. Power*, vol. 21, no. 2, pp. 1291–1309, 2011.

21 Y. Miao, Q. Jiang, and Y. Cao, "Battery switch station modeling and its economic evaluation in microgrid," in *2012 IEEE Power and Energy Society General Meeting*, San Diego, CA, pp. 1–7, 2012.

22 G. Cardoso, M. Stadler, A. Siddiqui, C. Marnay, N. DeForest, A. Barbosa-Póvoa, and P. Ferrão, "Microgrid reliability modeling and battery scheduling using stochastic linear programming," *Electr. Power Syst. Res.*, vol. 103, pp. 61–69, 2013.

23 W. Su, J. Wang, and J. Roh, "Stochastic energy scheduling in microgrids with intermittent renewable energy resources," *IEEE Trans. Smart Grid*, vol. 5, no. 4, pp. 1876–1883, 2014.

24 Z. Wu, W. Gu, R. Wang, X. Yuan, and W. Liu, "Economic optimal schedule of CHP microgrid system using chance constrained programming and particle swarm optimization," in *2011 IEEE Power and Energy Society General Meeting*, Detroit, MI, pp. 1–11, 2011.

25 Q. Sun, R. Han, H. Zhang, J. Zhou, and J.M. Guerrero, "A multiagent-based consensus algorithm for distributed coordinated control of distributed generators in the energy internet," *IEEE Trans. Smart Grid*, vol. 6, no. 6, pp. 3006–3019, 2015.

26 M. A. Hassan and M. A. Abido, "Optimal design of microgrids in autonomous and grid-connected modes using particle swarm optimization," *IEEE Trans. Power Electron.*, vol. 26, no. 3, pp. 755–769, 2011.

27 M. H. Albadi and E. F. El-Saadany, "A summary of demand response in electricity markets," *Electr. Power Syst. Res.*, vol. 78, no. 11, pp. 1989–1996, 2008.

28 L. Kristov and P. De Martini, "21st Century Electric Distribution System Operations," CAISO, 2014.

29 H. Kim and M. Thottan, "A two-stage market model for microgrid power transactions via aggregators," *Bell Labs Tech. J.*, vol. 16, no. 3, pp. 101–107, 2011.

30 E. Sortomme and M. A. El-Sharkawi, "Optimal charging strategies for unidirectional vehicle-to-grid," *IEEE Trans. Smart Grid*, vol. 2, no. 1, pp. 131–138, 2011.

31 A.-H. Mohsenian-Rad, V. W. S. Wong, J. Jatskevich, R. Schober, and A. Leon-Garcia, "Autonomous demand-side management based on game-theoretic energy consumption scheduling for the future smart grid," *IEEE Trans. Smart Grid*, vol. 1, no. 3, pp. 320–331, 2010.

32 L. Gkatzikis, I. Koutsopoulos, and T. Salonidis, "The role of aggregators in smart grid demand response markets," *IEEE J. Sel. Areas Commun.*, vol. 31, no. 7, pp. 1247–1257, 2013.

33 H. Zhong, L. Xie, and Q. Xia, "Coupon incentive-based demand response: theory and case study," *IEEE Trans. Power Syst.*, vol. 28, no. 2, pp. 1266–1276, 2013.

34 M. Parvania, M. Fotuhi-Firuzabad, and M. Shahidehpour, "Optimal demand response aggregation in wholesale electricity markets," *IEEE Trans. Smart Grid*, vol. 4, no. 4, pp. 1957–1965, 2013.

35 D. T. Nguyen and L. B. Le, "Risk-constrained profit maximization for microgrid aggregators with demand response," *IEEE Trans. Smart Grid*, vol. 6, no. 1, pp. 135–146, 2015.

36 S. Parhizi and A. Khodaei, "Market-based microgrid optimal scheduling," in *2015 IEEE International Conference on Smart Grid Communications*, Miami, FL, pp. 55–60, 2015.

37 E. Martinot, L. Kristov, and J.D. Erickson, "Distribution system planning and innovation for distributed energy futures," *Curr. Sustain. Energy Rep.*, vol. 2, no. 2, pp. 47–54, 2015.

38 New York State Department of Public Service Commission, "Developing the REV Market in New York: DPS Staff Straw Proposal on Track One Issues," New York State Department of Public Service Commission, 2014.

39 J. Taft and A. Becker-Dippmann, "Grid Architecture," PNNL, [Online]. Available: http://energy.gov/sites/prod/files/2015/04/f22/QERAnalysis-Grid Architecture_0. pdf, 2015.

40 F. Rahimi and S. Mokhtari, "From ISO to DSO: imagining new construct—an independent system operator for the distribution network," *Public Util. Fortn.*, vol. 152, no. 6, pp. 42–50, 2014.

41 Z. Wang, B. Chen, J. Wang, M. M. Begovic, and C. Chen, "Coordinated energy management of networked microgrids in distribution systems," *IEEE Trans. Smart Grid*, vol. 6, no. 1, pp. 45–53 2015.

42 P. De Martini, "More Than Smart: A Framework to Make the Distribution Grid More Open, Efficient and Resilient," Greentech Leadership Group, Aug. 2014.

43 J. Tong and J. Wellinghoff, "Rooftop parity: solar for everyone, including utilities," *Public Util. Fortn.*, vol. 152, no. 7, pp. 18–23, 2014.

44 S. Parhizi and A. Khodaei, "Investigating the necessity of distribution markets in accomodating high penetration microgrids," in *IEEE PES Transmission & Distribution Conference & Exposition*, Dallas, TX, 2016.

45 S. Parhizi, A. Khodaei, and S. Bahramirad, "Distribution market clearing and settelement," in *IEEE PES General Meeting*, Boston, MA, in press, 2016.

46 S. Bahramirad, A. Khodaei, and R. Masiello, "Distribution markets," *IEEE Power Eng. Mag.*, pp. 102–106, 2016.

47 A. Arab, A. Khodaei, and S. K. Khator, "Proactive recovery of electric power assets for resiliency enhancement," *IEEE Access*, vol. 3, pp. 99–109, 2015.

48 A. Khodaei, "Microgrid optimal scheduling with multi-period islanding constraints," *IEEE Trans. Power Syst.*, vol. 29, no. 3, pp. 1383–1392, 2014.

49 A. Khodaei, M. Shahidehpour, and S. Bahramirad, "SCUC with hourly demand response considering intertemporal load characteristics," *IEEE Trans. Smart Grid*, vol. 2, no. 2, pp. 564–571, 2011.

Index

Note: *Italicized* and **bold** page numbers refer to figures and tables, respectively.

The Economics of Microgrids, First Edition. Amin Khodaei and Ali Arabnya.
© 2024 The Institute of Electrical and Electronics Engineers, Inc.
Published 2024 by John Wiley & Sons, Inc.

Printed and bound by CPI Group (UK) Ltd, Croydon, CR0 4YY

16/04/2025

14658586-0001